Crystals
& Gems

Crystals
& Gems

Produced for DK by Toucan Books

DK LONDON
Senior Editor Scarlett O'Hara
Senior Designer Gillian Andrews
Senior US Editor Megan Douglass
Managing Editor Gareth Jones
Senior Managing Art Editor Lee Griffiths
Production Editor Robert Dunn
Production Controller Nancy-Jane Maum
Illustrator Emma Fraser Reid
Jacket Design Development Manager
Sophia M.T.T.
Jacket Designer Akiko Kato
Associate Publishing Director Liz Wheeler
Art Director Karen Self
Publishing Director Jonathan Metcalf

DK DELHI
Senior DTP Designer: Harish Aggarwal
Senior Jackets Coordinator: Priyanka Sharma-Saddi

First American Edition, 2023
Published in the United States by DK Publishing
1745 Broadway, 20th Floor, New York, NY 10019

A catalog record for this book
is available from the Library of Congress.
ISBN: 978-0-7440-8084-1

Printed and bound in China

For the curious
www.dk.com

FSC
www.fsc.org
MIX
Paper | Supporting
responsible forestry
FSC™ C018179

This book was made with Forest
Stewardship Council™ certified
paper—one small step in DK's
commitment to a sustainable future.
For more information go to
www.dk.com/our-green-pledge

Contents

Other Semiprecious Gems

Organic Gems

Introduction

Hewn from the Earth—or in the case of pearls and coral, plucked from the ocean floor—gemstones have entranced humans for millennia. They are expressions of love, symbols of wealth and power, and objects of beauty and adornment.

Ancient civilizations endowed gemstones with magical, talismanic powers. In Europe in the Middle Ages, they were linked to astrological signs and believed to provide protection and impart wisdom. Medieval apothecaries attributed healing powers to certain stones, just as crystal healers do today.

As trade widened around the world from the 15th century, gemstones were increasingly used as status symbols. Colombian emeralds were traded in Persia, and Indian sapphires were sold in Europe. As many of their portraits testify, monarchs and maharajas covered themselves in jewels, effectively flaunting their treasury.

This book includes the history, folklore, scandalous tales, and record-breaking prices associated with gemstones, from the four precious stones—diamonds, emeralds, rubies, and sapphires—to richly colored hard stones, such as carnelian, jade, and jasper. It includes spectacular jewels in the collections of Europe's royal families, and items that once belonged to well-known lovers of gemstones, such as Wallis Simpson, Elizabeth Taylor, and Elvis Presley. The pages feature exquisite pieces by the world's great jewelery houses and knock-out museum gems of unbelievable size and clarity.

Precious Gems

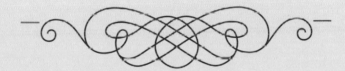

Diamond

Everlasting love

Diamonds are record breakers in many ways—the hardest natural material, the most brilliant gemstone, the most expensive stone—drawing a glittering constellation of superlatives. Coming from the depths of the Earth and formed over millions of years, diamonds are rich in symbolism, legend, and mythology. They have been linked to the gods and the stars, and are said to bring love, health, and wealth. But their power can have a dark side: some of the largest and most famous diamonds are said to be cursed.

All diamonds come from deep underground. Brought to the surface in kimberlite pipes—vertical columns of volcanic rock thrust to the surface at least 25 million years ago—diamonds were exposed by powerful geological forces, such as glaciation and erosion. Eventually they ended up in sandy riverbeds, where humans first discovered them. Left in the Earth, diamonds, like any mineral, are

" Better a diamond with a flaw than a pebble without. "

Confucius (551–479 BCE)

A brilliant-cut diamond

subject to pressures and heat that change them. In a billion or so years, diamonds would become a substance called graphite, a crystalline form of carbon used in lead pencils.

Humanity's love affair with diamonds began in India, where farmers first found diamonds in the soil left by floodwaters. In Hindu mythology, diamonds are called *vajras* (thunderbolts), and are said to have been created when lightning bolts hit rocks. Lord Indra, the

The affair of the necklace

In 1785, a supposed countess at the court of King Louis XVI convinced a gullible cardinal seeking royal approval to buy a diamond necklace for his queen, Marie Antoinette. But instead of delivering the necklace to the queen, the countess kept it. The trick was exposed when the cardinal failed to pay the jeweler, who then complained to the court. The cardinal was exiled and the countess fled to England, where she wrote an account of the affair criticizing Marie Antoinette. The queen was tried in the court of public opinion, found guilty of immorality, and eventually lost her head to the guillotine during the French Revolution.

Hindu god of rain and thunder, wields a thunderbolt, and diamonds, sometimes found in fields after heavy rain had washed away the clay soil, were thought to be gifts from the gods. They were used as currency and for their supposed power to heal body, mind, and spirit. They were also placed in the eye sockets of sculptures of the deities and worn by India's princes and generals to give them good health and strength in battle.

Indian diamonds reached the rest of the ancient world along trade routes or as war booty. Taken to Greece by Alexander the Great in 327 BCE, they were believed to be the tears of the gods or the splinters of stars. The Roman writer Pliny the Elder said in his *Natural History* (77 CE): "Diamond is the most valuable of all things in this world."

India's diamonds were later traded along the Silk Road to China in the east and Europe in the west. India's effective monopoly on diamonds lasted until 1725 when mines were established in Brazil. When the British conquered India in 1849, they took many ancient stones, including what was, at the time, the world's largest known diamond, the Koh-i-Noor (see below).

The jewel in the crown

The origins of the Koh-i-Noor Diamond, whose name means "River of Light" in Persian, are lost to the mists of time. Its first reliable appearance comes in 1628 when Mughal ruler Shah Jahan had it set into his jewel-encrusted peacock throne. More than a century later, Persian ruler Nader Shah took it when he sacked Delhi in 1739. The stone passed to the Durrani dynasty in Afghanistan before returning to India, where it was handed over to the British in 1849. Taken to England, it became part of the British Crown Jewels.

For Indians, the inclusion of the Koh-i-Noor in a British crown—currently the one that was worn by the late Queen Mother—is a reminder of Britain's colonization of their country. Calls for the stone to be repatriated, amplified after the death of Elizabeth II in 2022, have so far been resisted.

Cut, clarity, and color

With diamonds, the size, which is measured in carats, matters but so does the quality, or clarity, of the stone. Uncut, or "rough," diamonds are dull; jewelers cut them to reveal their sparkle. The hardest natural substance, a diamond can only be cut by another diamond. Diamond cutters divide the "rough" into stones, shaping each one to create the most valuable gem possible. The earliest faceted diamonds were "rose-cuts," but round "brilliant-cuts" are the most popular today because they best reflect the light. White diamonds range in hue from completely colorless, the most valuable, to pale yellow or brown. Fancy diamonds, colored by various elements, are far rarer.

White diamonds are the best known and among the most expensive stones, but diamonds come in all the colors of the rainbow—pink, red, blue, green, orange, violet, and yellow. In crystal healing, the color dictates their role. Yellow diamonds, for instance, may be placed on the solar plexus with the aim of benefiting organs located between the sternum and the navel, while white diamonds are believed to open the body's crown chakra, the energy point associated with the highest level of consciousness. Blue diamonds are thought to help unite mind and body.

> " My reason for choosing diamonds is that they represent the greatest worth in the smallest volume. "
>
> **Coco Chanel**, 1932

The most famous blue diamond is the Hope Diamond, a huge 45.52-carat stone said to be cursed. Certainly many of those who have owned it have suffered, most notably King Louis XVI and his queen, Marie Antoinette, who were executed during the French Revolution. The last woman to own the Hope Diamond, American heiress Evalyn Walsh McLean, adored jewels and had the money to indulge her passion. She fell in love with the priceless stone, which had been set by Cartier in a mount of white diamonds to enhance its blue color, later saying, "I put the chain around my neck and hooked my life to its destiny for good or evil." She let her Great Dane wear the diamond and hid it in her garden for her guests to play Find the Hope.

In the end, Evalyn had more than her share of bad luck: her husband left her, her daughter was killed in a car crash, and her son died of a drug overdose. After her own death in 1947, the stone eventually passed to the Smithsonian Institution, and can now be seen in the National Museum of Natural History, in Washington, D.C.

Hollywood star Elizabeth Taylor was another diamond obsessive, and the diamond given to her was almost more famous than she was. The huge diamond that graced her finger—and then her décolletage when she found it hard to lift her finger because of the stone's weight—was mined in South Africa by De Beers in 1966 and cut into a 60-carat pear-shaped stone by New York jeweler Harry Winston. The flawless stone came up for sale at auction in 1969

A tale of two engagement rings

When Monaco's Prince Rainier III asked American movie star Grace Kelly to marry him, he gave her a diamond and ruby eternity band. Created by Cartier in the colors of Monaco's flag and set with stones that were family heirlooms, the ring had sentimental value, but lacked star power so Rainier gave his bride-to-be another ring with more bling. Set with a central 10.48-carat diamond and two baguettes (long rectangular stones), it can be seen in her last film *High Society*.

Elizabeth Taylor wears the Taylor-Burton Diamond at the Oscars, 1968.

❝ Big girls need big diamonds. ❞

Elizabeth Taylor

> ❝ That diamond upon your finger, say
> How came it yours? ❞

William Shakespeare, *Cymbeline*, 1611

Beyoncé flashes the diamond engagement ring given to her by husband Jay-Z.

but the jeweler Cartier narrowly outbid Taylor's husband Richard Burton. Calling from a hotel in London, Burton bought it from Cartier the next day for $1 million and named it the Taylor-Burton Diamond.

While diamonds are said to be forever, love often is not: Taylor and Burton divorced in 1976. She sold the 20th-century's most famous love token in 1979 for $3 million and used part of the proceeds to fund a hospital in Namibia. The diamond's current owner is unknown.

No 21st-century engagement ring has so far exceeded the price of Grace Kelly's ring—around $4 million in 1956 and now valued at around $38.8 million—but the large sums paid for celebrities' rings regularly make news. The 35-carat ring designed by Wilfredo Rosado that Australian billionaire James Packer gave to Mariah Carey in 2016, for example, reportedly cost $10 million, making it a shock (and perhaps an insult) when the singer sold it for a mere $2 million after the romance ended. One of the most famous rings of recent years belongs to Beyoncé, whose husband Jay-Z gave her an 18-carat diamond designed by celebrity jeweler Lorraine Schwartz. In 2008, a year after their secret marriage, Beyoncé flashed the ring at the end of the video for her hit song "Single Ladies," in which she berates men who are unwilling to commit to their partners—men who had not, as she says in the song, "put a ring on it."

When advertising copywriter Frances Gerety wrote the slogan "A diamond is forever" for De Beers in 1947, with the aim of reversing the stone's declining sales, she made these sparklers the most coveted stone for engagement rings—not just for the rich, royal, and famous, but for everyone.

Blood diamonds

In 2021 alone, $13.99 billion of diamonds were exported from diamond-producing nations. But the supply line for diamonds can be murky, and reputable diamond traders have begun to ask questions about where this money is going. They want to ensure they are not dealing in "blood diamonds" (or conflict diamonds), defined by the United Nations as diamonds mined by forces that oppose their nation's legitimate government and sold to fund military action against it. Blood diamonds have been used to fund violent campaigns in Angola, the Democratic Republic of the Congo, and Sierra Leone. The 2006 film *Blood Diamonds* opened many people's eyes to the problem, but diamonds are still mined in war zones.

Emerald and diamond sarpech
(turban ornament), India, 18th century

Emerald

Life force

Grass-green, the color of new growth, emeralds have long symbolized renewal. In ancient Egypt, they were placed in the tombs of pharaohs to represent the annual flooding of the Nile River, which rejuvenated the parched lands that bordered it. The Roman writer Pliny the Elder, in his work *Natural History*, noted the refreshing effect of gazing on emeralds, how they soothed eyes weary from poring over other objects. An emerald worn around the neck was thought to guard against epileptic fits, to protect its wearer from evil spirits, and to blind serpents. Another tradition held that an emerald placed in the mouth would bring relief from dysentery. In modern crystal healing, emeralds are said to aid concentration.

Unlike many other gemstones, emeralds are treasured not so much for their sparkle or brilliance, as for the beauty and intensity of their color. Pliny said, "There is no stone, the color of which is more delightful to the eye." As Pliny recognized, emeralds are a form of the mineral beryl, and it is their chemistry that makes them so rare. The elements beryllium, aluminum, silicon, and oxygen that constitute

> ❝ Compare [the emerald] with
> other things, be they never
> so green, it surpasseth them
> all in pleasant verdure. ❞

Pliny the Elder, *Natural History*, c. 77 CE

beryl are not normally all found at the same level beneath the Earth's surface. Small traces of the elements chromium and vanadium provide impurities that give emeralds their green hue. Other forms of beryl contain different trace elements and include sea-green aquamarines, pinkish-red morganites, and yellow heliodors. Some of the gems Pliny called *smaragdus*, which is Latin for emerald, were probably different green stones, such as peridot or green sapphire.

In the ancient world, Egypt was the main source of emeralds. Archaeologists have uncovered the remains of emerald mines—often called Cleopatra's Mines, after the Egyptian queen, who was a lover of the stone—at Jabal Sukayt and Jabal Zabarah in southern Egypt's Eastern Desert. Worked from at least 1500 BCE, and probably much earlier, these mines supplied the Mediterranean area and South Asia. By today's standards, their stones would have been of inferior quality.

The Aztecs, Incas, and other Indigenous civilizations of the Americas knew and valued the stones of the highest quality. The Spanish conquistadors, Hernán Cortés in Mexico and Francisco Pizarro in Peru, and their followers, were entranced by the emeralds

An 18th-century cup set with emeralds and other gems, Mysore, India

The Crown of the Andes, dating from 1660 (diadem) and 1770 (arches)

they looted in their conquests. The sources of the stones remained a mystery—perhaps deliberately kept from them by the Indigenous people—until 1537, when the Spanish tracked down emerald mines in the eastern ranges of the Andes in modern-day Colombia. Enslaving the local people as miners, they sent emeralds flooding back across the Atlantic to Europe and beyond.

Emeralds have been found elsewhere, including Zambia, Ethiopia, Brazil, Russia, Austria, Australia, and North Carolina, but Colombian stones, prized for their incomparable intense blue-green color, still account for the bulk of the market. More than 400 emeralds adorn a

> ## " The ice, mast-high, came floating by, As green as emerald. "

Samuel Taylor Coleridge, *The Rime of the Ancient Mariner*, 1834

gold crown known as the Crown of the Andes, created for a sacred statue of the Virgin Mary in the cathedral of Popayán in Colombia around 1660, though it has had later enrichments. Its largest stone is called the Atahualpa Emerald, after the last Inca emperor. For centuries, the crown and its statue were carried in Holy Week processions in Popayán, but it was sold in 1936 to a group of American businessmen. Since 2015, it has been owned by the Metropolitan Museum of Art in New York City.

Among the most spectacular Colombian stones, and one of the largest uncut emeralds in the world, is the 1,383.93-carat Duke of Devonshire Emerald, believed to have been a gift to the sixth Duke of Devonshire in 1831 by Emperor Pedro I of Brazil, and now in London's Natural History Museum. As well as its size, it is

The "Mogul Mughal" Emerald

— ☾ ♠ ☽ —

Among the most spectacular and historic of Colombian stones is the so-called Mogul Mughal Emerald, one of the world's largest emeralds, now on display in the Museum of Islamic Art in Doha, Qatar. The rectangular-cut stone measures 2 in (5.2 cm) long by 1½ in (4 cm) wide, is ½ in (1.2 cm) thick, and weighs more than 217 carats. Engraved floral designs adorn one side of the stone (see left) and Islamic texts cover the other. The emerald is dated to the reign of India's sixth Mughal emperor, Aurangzeb (r. 1658–1707) and probably made its way from Colombia to India via European traders.

The Green Wars

Some 3,500 people died in the so-called Green Wars, which erupted in Colombia's Boyacá emerald-mining region in the 1980s. Tensions had developed in the 1960s when mine owners started to raise private armies to protect themselves from guerrillas and narcotics traffickers seeking a share of emerald export profits. Mafia-style feuding between two families, the Carranzas and the Rincons, who controlled most of the trade, added to the bloodshed. The violence eased in the 1990s, but flared up again in 2013, when the son of Pedro Rincon, head of the Rincon clan, was killed in a grenade attack. A growing global demand for luxury goods helps fuel this dark side to Colombia's emerald trade.

notable for its intense, luminous green. It also displays the natural hexagonal shape of uncut emeralds, and some of the host rock in which it was embedded is still visible at the base. The stone attracted crowds of admirers when it was put on show in London's Great Exhibition in 1851.

The world's largest flawless emerald is the Rockefeller Emerald, bought by John D. Rockefeller, Jr., in 1930 for his wife, Abby Aldrich Rockefeller. Originally part of a brooch, it now adorns a ring, which sold at auction for more than $5.5 million in 2017. Perhaps the most famous emeralds of the 20th century were those in an emerald and diamond parure created by the jewelers Bulgari in Rome. Consisting of a necklace, a brooch (which could also be worn as a pendant), earrings, a ring, and a bracelet, the jewels were bought by Richard Burton for Elizabeth Taylor, soon to become his second wife. At the time, the pair were in Italy filming the Hollywood extravaganza *Cleopatra* (1963). The jewels were among the most dazzling items in Taylor's famously brilliant collection.

The Rockefeller Emerald, set in a ring of white gold

" Blinded like serpents when they gaze
Upon the emerald's virgin blaze. "

Thomas Moore, *Lalla Rookh*, 1817

Ruby

The king of stones

" For wisdom is more
precious than rubies. "

The Bible, Proverbs 8:11

Known in ancient Sanskrit as *ratjaraj*, the "king of precious stones," this rare and fiery red gem is associated with power and wealth as well as with passion and romance. Today it is, carat for carat, the most valuable of all colored gemstones.

In ancient cultures, the ruby's blood-red color signified life. In Burma (now Myanmar)—the main source of rubies until the 20th century—rubies were known as "blood drops from the heart of Mother Earth." In many cultures, the stone was believed to protect the wearer from harm or misfortune. Burmese warriors, for example, believed that rubies provided a protective shield that made them invincible. They wore rubies on the left side of the body—the same side as the heart—and even had rubies inserted under the skin so that the gem became a part of their body. That way, they could not be wounded by "spear, sword, or gun." Ancient Chinese warriors wore rubies set into their helmets or armor for protection.

Indian dagger decorated with rubies and emeralds, c. 1605–1627

Rubies were also seen as empowering. In ancient India, Hindus believed that making a ceremonial offering of rubies to the god Krishna would lead to their rebirth as an emperor. The ancient

King James VI of Scotland and I of England (r. 1567–1625 and 1603–1625 respectively) wears the Mirror of Great Britain on his hat.

The Mirror of Great Britain

Commissioned by King James VI of Scotland and I of England to mark the Union of the English and Scottish crowns in 1603, the Mirror of Great Britain was composed of a large ruby, four diamonds (one suspended), and two large pearls. James had brought the gems to England from Scotland, where they had once formed part of the Great "H," a jewel belonging to Mary Queen of Scots. James's son, Charles I, later pawned the Mirror, along with other jewels, in order to raise money for his fight against Parliamentarians in the English Civil War (1642–1651).

Greeks believed that the rubies' fiery glow was proof that an inextinguishable flame burned within them, and that this would not only protect them when worn as an amulet, but could also bring a pan of water to a boil.

Ruby, like sapphire, is a version of the mineral corundum, which forms deep below Earth's surface through the combined forces of high temperatures and pressure. Traces of the element chromium give rubies their red color and fluorescence. Corundum is the hardest mineral after diamond, and it remains intact despite being transported by rivers from surface rocks to settle in gravel beds.

Rubies occur in colors ranging from orange or pinkish reds to a bluer, purplish shade, sometimes leading to confusion with other stones. Ancient Burmese miners thought that pink sapphires were "unripe" rubies. The most valuable rubies are an intense red, with depth of color and clarity, and good fluorescence. Some rubies display a luminous sheen created by tiny, needlelike inclusions known as "silk," which can lower their value. In some stones, however, the inclusions occur in a way that creates a cat's eye or star effect. Often cut cabochon-style

Ruby colors

The particular color of a ruby depends upon the type of rock in which it formed and the elements in its composition. Myanmar's rubies are chromium-rich, which gives them their bright red color. Sri Lanka produces pinkish or purplish reds with good fluorescence. Thailand's rubies have a higher iron content, resulting in a deep but less intense red due to less fluorescence, while those from Vietnam tend toward pink or purple. East African countries such as Tanzania, Mozambique, and Kenya produce good pigeon's blood rubies, as well as dark red and purplish stones. Tanzania also produces large opaque rubies that are used for carving.

to display the star to best effect, these rubies are highly valued. One of the largest and best star rubies is the 138.7-carat Rosser Reeves Ruby from Sri Lanka. American advertising executive Rosser Reeves was known to carry the stone, which was unset, around with him for good luck before giving it to the National Museum of Natural History, in Washington, D.C., in 1965. Another large star ruby, the six-pointed, 100.3-carat DeLong Star Ruby discovered in Burma in the 1930s, is now in the American Museum of Natural History in New York City.

High-quality rubies are found in Sri Lanka, Thailand, Mozambique, and Tanzania. Traditionally, however, the very best rubies have come from the Mogok region of Myanmar, a remote valley that is the source of many precious stones, including sapphire, spinel, topaz, and peridot.

In the 1960s, a boom period for Mogok, the 20-mile- (32-km-) long valley contained 1,600 small mines, which were jealously protected. At various times, the valley has been off-limits to foreigners, and today a

Pendant depicting a mythical Greek siren, c. 1580

special permit is needed to visit the area. The best Mogok rubies have an intense color, known as "pigeon's blood" red, and display strong fluorescence.

> **"** Those be rubies, fairy favors. **"**
>
> **William Shakespeare**, *A Midsummer Night's Dream*, 1605

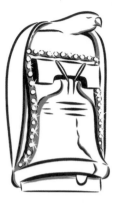

The Liberty Bell Ruby

To celebrate the US's Bicentennial in 1976, an 8,500-carat ruby—the largest mined ruby in the world—was carved into a tiny statue of the famous Liberty Bell in Philadelphia and decorated with 50 diamonds. In 2011, the piece was stolen from a secure store in Delaware. Four men were arrested for the theft in 2014, but the piece has never been recovered.

Very few pigeon's blood rubies are found in the Mogok area today, a rarity that drives up their value. Gem-quality rubies good enough for faceting are particularly unusual, especially in larger sizes (above two carats). The world's most expensive rubies include a Burmese pigeon's blood ruby, known as the Sunrise Ruby, set in a ring by Cartier. Bought by an anonymous Swiss bidder in 2015 for more than $30 million, at the time it was the most expensive faceted gemstone ever sold other than a number of famous diamonds. At 25.59 carats, it dwarfed the 8-carat Graff Ruby, the previous record-breaker, which had sold for just over $8 million at auction the previous year.

The Sunrise and the Graff rubies are in private collections, but the Carmen Lúcia Ruby, a 23-carat pigeon's blood ruby mined in the Mogok region in the 1930s, is on public display at the National Museum of Natural History in Washington, D.C. Set in a ring, between two triangular diamonds, the ruby was bought for the Smithsonian in 2004 by investor and physicist Peter Buck in memory of his late wife, after whom the stone is named. It is one of the largest faceted rubies in the world.

Rare stones such as the Sunrise often come onto the market after spending many years hidden away in family vaults. One of the world's more mysterious rubies was revealed to the public in 1986 and then locked away in a secret vault. Known as the Rajaratna Ruby, it is owned by Mr. G. Vidyaraj, a retired solicitor living in Bangalore, India, who inherited a collection of rubies that had

The 25.59 Sunrise Ruby (right) fetched over $30 million at auction in 2015, about $9 million more than this 100.20 carat diamond (left).

> ❝ Enter a mine of rubies and bathe
> in the splendor of your own light. ❞

Rumi, 13th century

purportedly been in his family for centuries. Wanting to find out about the gems in his collection, Vidyaraj took a home-study course in gemology and realized that one of his stones was a rare star ruby.

Flawless rubies are unusual, and natural rubies can be treated to improve their color and clarity. Heat can be applied to remove bluish tones, or to dissolve silk in order to improve a stone's transparency. Small fractures can be treated with fillers or healed with heat treatments. Such enhancements must always be declared.

The first synthetic rubies were produced around 1885 in Switzerland and were known as Geneva Rubies. They consisted of fragments of natural ruby fused together. Around the same time, French chemists Edmond Frémy and Auguste Verneuil were experimenting with a process of heating alumina and chromium oxide, which produced droplets of bright red crystal that could be polished and cut to produce beautiful synthetic rubies. Modern synthetic rubies are real rubies grown in a laboratory in a fraction of the time that natural rubies take to form. One of the main differences between natural and synthetic versions, other than the price, is the absence of inclusions in synthetic rubies.

The ruby's association with romance and passion, as well as its beautiful color, make the stone a popular choice for engagement rings. The most common cuts are ovals and cushions, which show off its crystal structure. Other shapes are rare in larger rubies.

Rubies continue to feature in the jewelery collections of the rich and famous. One of the most famous pieces of jewelery in the world is the ruby and diamond Cartier necklace, featuring eight cushion-shaped and oval-faceted Burmese rubies, that was given to the actor Elizabeth Taylor by her third husband, film producer Mike Todd, as she finished a swim one day. "I mean, how many young women get a set of rubies just for doing something wholesome like swimming laps? ... Well, I did," Taylor said. She wore the necklace with matching earrings and a red Dior dress to the 1957 London premiere of *Around the World in Eighty Days*,

The healing ruby

American mineralogist George F. Kunz wrote in *The Curious Lore of Precious Stones* (1913) that rubies were thought to remedy "hemorrhages of all kinds, as well as inflammatory diseases, [and] to exercise a calming influence." In modern crystal healing, rubies are said to stimulate the heart chakra, balancing the body and the emotions; sharpen the mind; and protect against negative forces. They are also believed to detoxify the blood and treat infectious diseases.

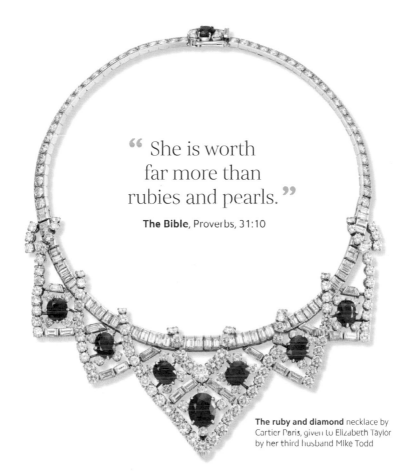

" She is worth
far more than
rubies and pearls. "

The Bible, Proverbs, 31:10

The ruby and diamond necklace by
Cartier Paris, given to Elizabeth Taylor
by her third husband Mike Todd

for which Todd won a Golden Globe Award for Best Picture. The
Associated Press described the stunning effect of Taylor's ensemble:
"Yards of crimson chiffon trailed from her sun-bronzed shoulders
to the floor. Her ruby and diamond earrings and necklace glittered."
When Taylor's extensive jewelery collection was auctioned
after her death in 2011, the ruby and diamond necklace alone
fetched $3,778,500.

Sapphire

Infinite pools

Maria Feodorovna, Empress of Russia (1881–1894),
wearing her sapphire parure—a matching set of jewelery

> **"** Silence was pleased: now glowed the firmament With living sapphires. **"**

John Milton, *Paradise Lost*, 1667

The cornflower blue of the most prized sapphires evokes the sky or the ocean, something pure blue and immense stretching toward infinity. The stone was associated with the heavens in South and Southeast Asian cultures; in medieval Europe, it symbolized purity, a detachment from earthly things. Sapphires often adorned bishops' rings, indicating their wearer's dedication to the divine.

Sapphires were also believed to have healing properties. One medieval English treatise describes the stone's "vertue" as "contrary to venym [poison]." Sapphires were also said to be an effective treatment for eye ailments.

Like rubies, sapphires are a kind of corundum, a crystalline form of aluminum oxide. Pure corundum is colorless, but this is rare. More often it is colored by minute traces of metals. Rubies are colored red by traces of chromium; all other corundum are classified as sapphires and can be green, yellow, or pink, as well as blue, the most prized, depending on the traces that are present. The second hardest natural substance after diamond, corundum formed under colossal pressure and at massively high temperatures deep in the Earth. Over millions of years, some corundum rose to the surface embedded in volcanic rock; separated out by water and weathering, it was then borne down mountainsides by streams and deposited in rivers.

Sapphires are found in parts of Africa, Madagascar, Australia, North America, and Asia, but Sri Lanka, Myanmar,

The Logan Sapphire, a 423-carat stone from Sri Lanka

(Burma), and the valleys and floodplains of Kashmir have yielded up the world's most treasured stones. In Sri Lanka, they have been mined since at least the second century CE, when the Egyptian astronomer Ptolemy referred to the island's sapphires, as did the Italian traveler Marco Polo in the 13th century.

Sri Lanka is also (along with Tanzania) a source of extremely rare "color change" sapphires, in which certain combinations of metal traces cause their color to change depending on the light source. When the light is rich in blue wavelengths, as in daylight, such stones are an intense blue. In other lights, the same stones will appear to have a redder or purplish hue.

Kashmir sapphires, in particular, set collectors' pulses racing. Many of the finest specimens were mined during a period of just over seven years in the 1880s, adding rarity value to their brilliant, satiny blue. Very few Kashmir sapphires are found outside museums, but in 2015, records were broken at an auction at Sotheby's in Hong Kong when a 27.68-carat Kashmir sapphire set with small diamonds and mounted

Pearl and sapphire earring, 7th century

Star sapphires

In some sapphires, the presence in the gem of minute needlelike structures of rutile (titanium dioxide) creates an optical effect called "asterism"—when illuminated, a six-rayed star appears to shine within the stone. Such stones, milky rather than translucent, are shaped and polished en cabochon to create a dome-like upper side, in which the star is centered, and a flat base. Cabochon-cut star stones include the largest sapphires ever mined, such as the 1,404-carat Star of Adam stone found in Ratnapura, Sri Lanka, in 2015, and the spectacular 563-carat Star of India, also from Sri Lanka, donated to the American Museum of Natural History in New York City by the financier J. P. Morgan. In 1972, Richard Burton gave a cabochon sapphire to Elizabeth Taylor for her 40th birthday.

on a white gold ring, was snapped up by a private collector for more than $6.7 million. At the time, this was the highest price ever paid per carat for a sapphire.

Royalty and nobility have been among the greatest lovers of the blue gemstone. In 1796, the young Napoleon, then a French general, gave his fiancée, Joséphine de Beauharnais, an engagement ring mounted with two pear-shaped jewels set side by side: a diamond and a sapphire. Two historic sapphires adorn the Imperial State Crown,

> **"... the open sky sits upon our senses like a sapphire crown "**
>
> John Keats, Letter to Jane Reynolds, 1817

The Palatine Crown (c. 1370–1380), thought to have been worn by Anne of Bohemia, wife of King Richard II of England, contains pale blue sapphires. It is now in the Munich Residence, Germany.

part of the British Crown Jewels, seen by millions watching the funeral of Queen Elizabeth II in 2022 when it rested on top of her coffin. At the center of the cross surmounting the crown is St. Edward's Sapphire, the oldest gemstone in the royal collection, believed to have been mounted on a ring worn by the 11th-century English king, Edward the Confessor, who was also venerated as a saint.

> **"** Sapphire feels the air and sympathizes with the heavens. **"**
>
> **Gaius Julius Solinus**, 3rd century CE

Made for French socialite Daisy Fellowes by Cartier Paris in 1936, and altered in 1963, the "Hindu" necklace contains sapphires, diamonds, emeralds, and rubies.

Lady Diana's ring

Famously, Lady Diana Spencer chose a sapphire for her ring when she became engaged to Britain's Prince Charles (now King Charles III) in 1981. She is said to have broken with tradition by choosing a 12-carat oval sapphire from London jewelers Asprey instead of selecting a stone from the royal collection. In 2010, 13 years after Diana's death, her elder son, Prince William, gave the same ring to his fiancée, Catherine Middleton.

The Stuart Sapphire, at the base of the crown, is thought to have belonged to Charles II, of the Stuart dynasty. When his brother and successor, James II, was forced to flee to France during the Glorious Revolution of 1688, he is believed to have taken the Stuart Sapphire with him. After various vicissitudes, the stone eventually found its way to the Prince Regent, later George IV, who gave it to one of his mistresses. It was reclaimed for the royal collection in time for the coronation of Queen Victoria in 1838.

In the 20th century, the workshops of the jeweler Cartier have been associated with some of the most spectacular sapphire settings. These include the pendant with a huge 478-carat stone—one of the world's largest faceted sapphires—owned by Queen Marie of Romania, by birth a British princess, and worn by her at the coronation of her husband, King Ferdinand, in 1922. Another Cartier design—on display in the National Museum of Natural History in Washington, D.C.—is the Bismarck Sapphire Necklace, created for the American-born socialite, Countess Mona von Bismarck in 1959, and later gifted to the Smithsonian.

Cartier also began combining carved jewels in extravagant combinations. The Art Deco Indian-inspired "Hindu" necklace belonged to what later became known as Cartier's Tutti-Frutti collection. It includes 13 briolette-cut sapphires weighing 146.90 carats in total, two leaf-shaped carved sapphires (50.80 and 42.45 carats), sapphire beads, and one sapphire cabochon.

Quartz and Chalcedony

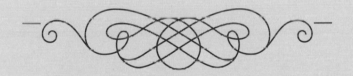

Rock crystal sculpture of the Buddha, Nepal, 17th century

Rock Crystal

Pure magic

The purest of all the varieties of quartz, rock crystal is associated with new beginnings. In crystal healing, it is believed to offer spiritual purification by cleansing all of the body's seven energy points (chakras), and also to strengthen the immune system.

Rock crystal contains no color-causing trace elements. Before glass was invented, it was highly prized by ancient cultures for its transparency—no other readily available material had the same qualities. The ancient Egyptians carved rock crystal into small

> ❝ Crystal is only to be found in places where the winter snow freezes with the greatest intensity; and it is from the certainty that it is a kind of ice. ❞

Pliny the Elder, (c. 77 CE)

figurines, vessels, and beads for jewelry, and set it in the eyes of carved limestone and wooden statues for an especially lifelike effect. Rock crystal was also used for ceremonial purposes at Neolithic (6000–3000 BCE) burial sites. In 2022, archaeologists at Dorstone Hill, an ancient burial site on England's border with Wales, found small pieces of rock crystal, probably brought there from more than 80 miles (130 km) away, scattered over cremated bone fragments.

The ancient Greeks believed that rock crystal was ice. In fact, quartz is present in all kinds of rocks. Forming in hot, watery solutions in veins or pockets in rocks, as well as in magma, crystals range in size from microscopic (cryptocrystalline) to

Gold and rock crystal
pendant, 4th century CE

> " What Youth deemed crystal,
> Age finds out was dew. "

Robert Browning, *Jocoseria*, 1883

enormous (macrocrystalline) and are instantly recognizable by their pyramidal ends, though the base is often broken where the crystals have been removed from the surrounding rock.

Rock crystal enjoyed widespread popularity during the Art Deco period in the early 20th century. Prized for its icy appearance, it was highly polished or frosted and often used as an engraved backdrop in rings, brooches, chokers, and necklaces. It was frequently set with other gemstones, such as white diamonds. It provided a striking contrast to black onyx and was also carved into three-dimensional, abstract, geometric shapes, reflecting the architectural and industrial styles of the period. Larger crystals were sometimes fashioned into a bracelet band with a gem-set platinum clasp.

Cartier embraced rock crystal not only for use in its jewelry, but also for one of its most elegant creations: the Cartier Mystery Clock, invented in the early 20th century by Louis Cartier and

Crystal balls

Said by crystal healers to help amplify and manifest a person's innermost desires and dreams, and to attract love, rock crystal has long been fashioned into crystal balls, in which fortune tellers claim to conjure visions of the future. Scrying, or crystal gazing, was common in Roman times—Pliny the Elder describes soothsayers using crystal balls—but was condemned by the early Christian Church. In Victorian times, crystal gazing became a popular pastime.

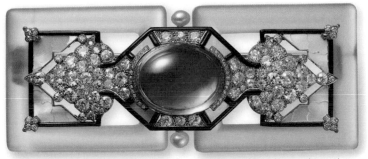

Sapphire, diamond, natural pearl, mother-of-pearl, and enamel brooch in a rock crystal setting, Cartier Paris, 1924

master horologist Maurice Couët and inspired by French illusionist Jean-Eugène Robert-Houdin. The clock hands are attached to disks of rock crystal and appear to be floating in mid-air, as if enchanted by a magician. The secret to the mechanism—hidden in the base of the clock—was fiercely guarded. The first generation of Mystery Clocks, known as "Model A," was released in 1912, and one of the first examples was bought by American financier John Pierpont Morgan.

Rock crystal also has a remarkable hidden property: piezoelectricity. This means that when a piece of rock crystal is placed under pressure, one end generates a positive charge and the opposite end a negative charge, making it able to support an electric current. The small pieces of quartz used in quartz timepieces vibrate at a precise rate of 32, 768 pulses a second when an electric current is applied to the quartz from a battery. This generates one electrical pulse per second, enabling the watch to keep impeccable time.

In 1969, the Japanese watch company Seiko created and marketed the first ever quartz watch, the Seiko Quartz Astron, with a price tag of 450,000 yen ($1,250), which was roughly the same as a small car. Requiring fewer components and said to be a hundred times more accurate than a standard mechanical watch, the Astron revolutionized watchmaking, paving the way for the first digital watches ten years later.

Amethyst

Deep purple

Favored for its purple hue, amethyst is the most popular variety of quartz and the best known purple gemstone. The ancient Greeks believed that it helped maintain sobriety if it was worn, held in the mouth, or used as a goblet for drinking alcohol. Its name is derived from the ancient Greek word *amethystos*, meaning "not drunken."

> " Amethyst dissipates evil thoughts and quickens the intelligence. "
>
> **Leonardo da Vinci**

Until the 18th century, when Portuguese colonists became aware of large deposits of amethyst in Brazil, the gemstone ranked alongside diamond, sapphire, ruby, and emerald in terms of value. Worn as beads or intaglios (with designs carved into the stone), it was shaped and faceted to display the prized color, which ranges from pale lilac to deep purple with flashes of red. Worn as an amulet, amethyst was believed to be an antidote to poison and to protect warriors in battle.

Amethyst typically forms in pockets of rocks known as geodes. The best amethyst comes from

Amethyst geode

An amethyst forms the center of a necklace (late 2nd–1st centuries CE) found in a tomb in Ukraine in 1891.

Uruguay, Zambia, Brazil, Tanzania, Russia, and the US, but it is found worldwide. Catherine the Great reportedly had thousands of miners searching for the gemstone in the Ural Mountains, and the highest grade of amethyst—an intense purple, mixed with some blue and with flashes of red—is sometimes referred to as Siberian.

Geodes can be used as decorations in their own right: they are miniature caverns filled with thousands of pointed purple crystals. The largest amethyst geode ever found was discovered in Uruguay's Artigas mine in 2007. Standing at an incredible 11 ft (3.4 m) tall, the Empress of Uruguay as it is known, is filled with tens of thousands of gem-quality amethysts of a saturated, deep purple. The geode was purchased by Australia's The Crystal Caves museum, which shipped the 2½-ton- (2.3-metric-ton-) geode (which is nearly as heavy as an elephant) 9,100 miles (14,645 km) to the Queensland museum.

Amethysts have long been associated with Christianity, their purple color symbolizing purity of spirit, and they are often set into the finger rings of bishops. Because of its purple color, amethyst is often used to represent grapes. The Cheapside Hoard, a collection of 16th- and 17th-century jewelery found in London

> ❝ Do you think amethysts can
> be the souls of good violets? ❞

L. M. Montgomery, *Anne of Green Gables*, 1908

in 1912, includes pendants of small clusters of grapes carved in amethyst. Later examples include an exquisitely decorated panel in the Hamilton Palace Cabinet (1844), now in London's Victoria & Albert Museum, featuring a bird feeding on grapes created from round amethyst cabochons.

Amethysts can be seen among the jewels of many royal families. In the scepter of the British regalia, a faceted amethyst *monde* (orb) sits above the Star of Africa Diamond that was presented to King Edward VII by the British colonial government in South Africa in 1907. One of the most impressive amethyst demi-parures, including a tiara, two bracelets, a *devant de corsage* (large triangular brooch) and drop earrings, belonging to the Swedish royal family, was once the property of Joséphine de Beauharnais, Napoleon Bonaparte's first wife. The tiara, converted from a necklace in the 1970s, consists of 15 faceted amethysts set in gold, surrounded by diamonds set in silver. Perhaps the most famous piece of amethyst jewelery is the amethyst, turquoise, and diamond bib necklace created by Cartier for the Duchess of Windsor in 1947, featuring 27 step-cut amethysts and a central heart-shaped amethyst. The duchess debuted the necklace at a gala ball in the Palace of Versailles in 1953— a dazzling piece of statement jewelery set off by a strapless white gown.

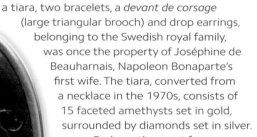

Roman intaglio, dating from
1st–2nd centuries CE

QUARTZ AND CHALCEDONY

46

The Duchess of Windsor wears her amethyst, turquoise, and diamond bib necklace, 1953

Love tokens

Amethyst is a staple of acrostic jewelery, in which the particular combination of gemstones conveys a secret message. A French import to England in the 18th century, the craze for acrostic jewelery reached its peak in the Victorian era. Gemstones representing different letters were arranged to spell words such as "REGARD" (R for ruby, E for emerald, G for garnet, A for amethyst, R for ruby, and D for diamond), "DEAREST," or "ADORE." The stones had differing meanings and amethyst stood for honesty and modesty, a chaste form of love. Amethyst was also associated with Valentine's Day. St. Valentine reputedly wore a ring set with an amethyst intaglio engraved with an image of Cupid, the Roman god of love.

Rose Quartz

The flush of youth

C rystal lore has appropriated the Greek myth of Aphrodite, Greek goddess of love, and Adonis, a handsome youth, to explain the color of rose quartz. According to its version of the myth, after Aphrodite abandons her previous lover, Ares, god of war and courage, for Adonis, and Ares mortally wounds the youth, her blood mixes with that of Adonis when she cuts herself on a briar bush, staining white quartz pink. Taking pity on the lovers, Zeus, the chief deity and father of Aphrodite, brings Adonis back to life for six months of every year.

> " Though we cannot understand the life of a crystal, it is nonetheless a living being. "

Nikola Tesla, 1900

The story has contributed to rose quartz becoming a symbol for reconciliation and undying love for those who believe in the supernatural power of crystals. It is said to heal emotional wounds, dissipate fear, and promote love for oneself and for others. It is also associated with eternal youth.

Rose quartz is abundant in many parts of the world, but the best quality stones come from Brazilian mines. Coarse-grained and often found in massive form (a mass of tiny crystals that cannot be seen individually), it can be fashioned into any shape, from a small bead to a large bowl or sculpture. In the 19th century, large quantities of rose quartz were imported into Imperial China, where they were turned into luxury items, from carved snuff bottles to elaborate representations of Chinese myths.

The cloudy appearance of most rose quartz means that it is rarely faceted (cut with flat angled surfaces to reflect light). Rose quartz owes its pink color to myriad inclusions of the mineral dumortierite, and sometimes displays asterism—a six-rayed starry effect that is achieved when light reflects off the inclusions. One of the largest star rose quartzes, in the collection of Michael Scott, the first president of Apple Inc., is 5,500 carats and fashioned into a sphere with a perfectly centered six-rayed star.

Snuff bottle, China, c. 1880–1925

Bird-on-a-rock, a citrine and diamond brooch by Jean Schlumberger for Tiffany, 1995

Citrine

Money and happiness

" Comets, importing change
of times and states,
Brandish your crystal
tresses in the sky. "

William Shakespeare, *Henry VI, Part I,* c. 1591

Golden-yellow or orange, citrine takes its name from *citrus*, the Latin word for lemon. However, most of the citrine on the market today probably began in Brazil as purple amethyst quartz before being heated to create its characteristic sunny yellows and oranges. Naturally occurring citrine is found in Madagascar and the Ural mountains in Russia, but it is very rare.

Believed to bring prosperity and abundance to the wearer, citrine is associated with success in business. It was a sought-after gem in the Greco-Roman period (325 BCE–395 CE), when it was used for charms and amulets, sometimes engraved with the figure of divine success, known as Bonus Eventus (Good Outcome). Such intaglios were also popular in the early 19th century.

Because citrine can be faceted into gemstones of considerable size, it is ideal for cocktail rings and statement jewelery. In the 1960s and '70s, Anglo-Italian celebrity jeweler Andrew Grima created rings, brooches, and bracelets set with larger-than-life citrines that captured the flamboyance of the times. Sometimes they were embellished with diamonds, and set on a background of highly textured gold. The warmth of the citrine was echoed by the yellow gold, a gorgeous antidote to post-war austerity.

Cairngorm brooches

In the 19th century, Queen Victoria popularized citrine through her love of Scotland, where the gemstone often formed the center of the ring brooches used to secure tartan plaid at the shoulder. Known as Cairngorm brooches, because citrine is found in Scotland's Cairngorm mountains, they became popular among tourists who wanted to buy souvenirs that evoked the colors of the Scottish countryside. Citrine gemstones were also set in dagger hilts and in snuff mulls (containers for powdered tobacco). Some snuff mulls were made from a ram's horn, or even an entire ram's head, and inlaid with a lidded silver compartment set with citrine.

Other companies also used citrine to great effect during the mid-20th century. American jeweler Tiffany & Co. used it in its bird-on-a-rock brooches, designed by Jean Schlumberger in the 1960s, when the company began to embrace the potential of lesser-known, colored gemstones. In Paris, esteemed jewelery designer Joel Arthur Rosenthal also made use of citrine. A pair of yellow pansy brooches, designed in 1988 for the late Ann Getty, wife of oil tycoon Gordon Getty, utilized the gradations in citrine's color to create a sense of depth.

Citrine and topaz are visually very similar, and citrine is sometimes referred to as "Spanish topaz" or "golden topaz." But there are differences between the stones. Their luster—the way light reflects off the surface—differs. Topaz is a harder mineral, which gives it a higher or brighter luster. Being softer, citrine can be carved into small figurines and sculptures. German gem carver Patrick Dreher, who belongs to a long dynasty of famous hard stone carvers, has used citrine for a number of his stunning animal sculptures. His citrine hippopotamus (2015) looks supple, yet rubbery, as a hippo should.

In crystal healing, citrine is sometimes known as the happy stone and is said to foster optimism, clearing the mind of negative thoughts and instilling calm. Its reputation for being the "money

Ametrine

Citrine forms an integral part of a bicolored variety of quartz known as ametrine. Its name combines "ame" from amethyst and "trine" from citrine. Typically, the gemstone is cut with an equal division of golden-yellow citrine and dark purple amethyst, with a distinct line down the middle. Both traditional gem cuts and more elaborate fantasy cuts are used to fashion ametrine into colorful and exciting pieces of wearable art. Ametrine rarely occurs in nature, and its main commercial source is the Anahí mine in southeastern Bolivia. More often, amethyst is partially heated to create yellow and purple geometric patterns.

Moth pendant
incorporating citrine
gold, and carved
horn, c. 1900.

" Citrine promotes inner calm
so that wisdom can emerge. "

Judy Hall, *The Crystal Bible*, 2003

stone," probably on account of its rich golden color, still remains.
Advocates of the ancient Chinese theory of feng shui—the
arrangement of a living space in order to optimize the balance of
energy, and therefore the spiritual well being of its inhabitants—
sometimes recommend placing a small tree decorated with citrine
in the home to attract good fortune.

Prasiolite

Perfect clarity

P rasiolite is a pale green variety of quartz. Its restful color and wondrous transparency make it a popular healing crystal. It is associated with the crown chakra (on the top of the head) and purported to help calm the mind, balance the emotions with reason, and enhance memory and logic. Worn as jewelery, it is said to boost courage and stimulate creativity.

Small quantities of naturally occurring prasiolite are found in Brazil, Poland, and Canada, but it is extremely rare. It forms in amethyst-bearing rock

Prasiolite and diamond
earrings, c. 2000

that has been heated by younger lava flows. The majority of prasiolite on the market is actually amethyst that has been irradiated to produce a pale green identical to that of its naturally occurring counterpart or that has been artificially heated to produce a stone that is a darker green. For this reason, prasiolite is sometimes marketed as "green amethyst," a description that is illegal in the United States.

Prasiolite is sometimes used as an inexpensive substitute for emerald or green-blue aquamarine. Relatively free of inclusions, it can be faceted to show off the clear green color, usually as a step-cut or rose-cut stone, or can be cut and polished into beads. It is a popular stone for pendant earrings and cocktail rings. However, the color of artificially produced prasiolite is unstable and can fade over time, especially when it is regularly exposed to the ultraviolet rays present in sunlight.

Aventurine

All that glitters

The name aventurine comes from *a ventura*, the Italian word for "chance," owing in part, as the story goes, to a mishap in an 18th-century Italian glass workshop. One day, copper filings were accidentally tipped into a batch of molten glass. When the glass cooled, light reflected off the copper filings, producing a sparkling effect that became known as aventurescence. The name was eventually bestowed upon aventurine quartz because of its similar glittering appearance.

Slightly translucent or opaque, aventurine typically occurs in shades of green, but can also come in reddish brown, pink, yellow, or blue. The stone is composed of colorless quartzite that has many inclusions, such as flakes of fuchsite (which causes green), a lithium-rich mica (creating pink), or an iron oxide (producing a reddish brown). When most of the inclusions are orientated in one direction, the sparkling aventurescence is displayed. If a stone contains more than one kind of inclusion, its color can be uneven.

The stone, which is found in several countries, including Russia, Germany, South Africa, India, and the US, is fashioned into beads and cabochons for rings and pendants, and carved into ornaments. Green aventurine is sometimes sold as a jadeite imitation.

Believers in the mystical powers of crystals consider aventurine to be a stone of good fortune, especially green aventurine, which is thought to bring good luck. Green aventurine is also believed to have soothing qualities, and be able to calm the mind and heal a broken heart.

Pink aventurine and green nephrite snail by Fabergé, Russia, c. 1885–1905

55

Smoky Quartz

———◇———

Dark art

Sometimes compared to a midnight cloud beneath a full moon, smoky quartz is the dark and mysterious cousin to rock crystal quartz. It can be light to dark brown, or even black in color, and transparent or opaque. The smoky quartz that is fashioned into jewelery, figurines, bottles, and bowls is usually transparent and lacks inclusions that are visible to the naked eye.

Found around the world as single crystals that range in size from tiny to huge, smoky quartz has been used for adornment for thousands of years. It was considered a powerful and sacred stone by Celtic

Smoky quartz snuff bottle, China, late 18th century

> " In the realm of dark stones,
> smoky quartz offers a rare
> and beautiful transparency. "

Crystal Vaults

tribes who inhabited parts of Europe from the 2nd millennium BCE. In medieval Europe, smoky quartz was often among the gemstones used to decorate ecclesiastical objects, such as chalices, altar screens, and crosses.

Like citrine, smoky quartz is found in abundance in Scotland's Cairngorm mountains, and helped spur a craze for gem collecting in the area in the mid-19th century. In 1851, a local man called James Grant presented Queen Victoria with a 50 lb (23 kg) chunk of smoky quartz called the Cairngorm Stone while she was staying at nearby Balmoral Castle. Traditionally, smoky quartz was set into the handle of the *sgian-dubh* (a small knife that kilt-wearing

> " Fabergé chose
> his stones
> to imitate the
> natural coloring
> of an animal. "

Géza von Habsburg, 1987

Fabergé bulldog, c. 1895–1915,
carved from smoky quartz

men tucked into the outward side of a sock), befitting its reputation as a protective stone.

In China, smoky quartz was used to fashion snuff bottles during the Qing dynasty (1644–1912), when snuff (powdered tobacco) became popular. Painstakingly hollowed out and carved with depictions of people, animals, or decorative motifs such as scrolls, or left smoothly polished, these snuff bottles were symbols of wealth and status. The carver would sometimes utilize variegated coloring within the stone to evoke images. If the smoky quartz was light enough in color, scenes and words were sometimes painted or inscribed on the inside of the vessel, where they would not be eroded by external wear and tear. The stoppers were typically made from glass or set with a contrasting gemstone, such as pink tourmaline.

In 19th-century Russia, smoky quartz was used by Imperial jeweler Carl Fabergé to create *bonbonnières* (dishes for holding confectionery) in the form of seashells with intricately enameled lids attached with gold hinges. Fabergé also created a collection of realistic animal and bird figurines inspired by Japanese *netsuke* (carved toggles attached to the sashes of kimonos) from a variety of hard stones, including smoky quartz, mined in Russia's Ural Mountains. Many of these animal figures, which included eagles, mice, elephants, dogs, and rabbits, were carved in the German town of Idar-Oberstein, a major center for Germany's gemstone industry, and famous for the quality of its hardstone carving. The figures were then returned to the Fabergé workshop in St. Petersburg to be embellished with gold details, such as feet and tails, and perhaps set with gems. Diamonds were often used as eyes. Such animal carvings were popular in the first decade of the 20th century, and were exported around the world. In Britain, King Edward VII and Queen Alexandra bought hundreds of them in a variety of hard stones, and commissioned sculptures of their favorite dogs and horses.

Smoky quartz gets its color from natural radiation, but rock crystal quartz can also be irradiated artificially to produce a smoky brown to black color. The color can then be lightened through heating, either in a laboratory or by ultraviolet rays over time. When heated, the color can take on a yellow hue, effectively turning smoky quartz into citrine, a more valuable stone. This lightening can be reversed by a further course of radiation.

Crystal healers believe they can dispel negative energy by pointing a piece of smoky quartz away from the body and drawing the energy through the stone. Considered to be a pragmatic stone, smoky quartz is said to help increase concentration, resolve contradictions, and relieve minor pain such as headaches.

The first sunglasses

In 12th-century China, panes of smoky quartz were held up to the eye to reduce the glare of the sun. Known as *ai-tai* ("dark clouds covering the sun"), such lenses were later incorporated into a metal frame with bending "arms" that could be secured around the ears, effectively becoming the first sunglasses, although the lenses did not protect the eyes from ultraviolet rays. However, official documents reveal a secondary purpose: such glasses were sometimes worn by judges to help conceal their expressions in court.

Carnelian

The lion stone

Dark red to reddish orange, carnelian's blazing colors associate the gem with the power of the sun and the courage of lions. Faceted carnelian beads dating from the middle of the 5th millennium BCE were discovered in the Varna Chalcolithic Necropolis, Bulgaria, when it was excavated in the 1970s. Each of these elongated cylindrical beads features 32 facets, the earliest type of complex faceting on a hard stone, such as quartz, ever recorded.

Ancient Egyptians used carnelian for beads, small figurines, and amulets, frequently in combination with lapis lazuli and turquoise. The stone was associated with the sun god, Ra, and sun disks, emblematic of Ra, were often carved from carnelian. The stone was also used to represent the eye of Horus, the falcon-headed god of light, and Apis, the bull deity. The ancient Egyptians believed that carnelian protected the wearer from falling stones, and master architects often wore it for that reason as well as to denote their status.

" Every kind of precious stone covered you: carnelian, topaz, and diamond, beryl, onyx, and jasper, sapphire, turquoise, and emerald. "

The Bible, Book of Ezekiel, 28:13

Cabochon carnelians decorate a pendant from Iran or Central Asia, mid-late 19th century

The ancient Romans believed carnelian would bring prosperity. Small pieces were carved with intricate images of emperors, gods, and goddesses, and set into rings, insignia, and other forms of adornment. These small carvings became popular again during the Italian Renaissance (14th–17th centuries), when the rediscovery of ancient Roman and Greek sites contributed to the revival of hard stone engraving. Similarly, new archaeological discoveries in the Mediterranean and Near East in the 18th and 19th centuries led to a passion in the West for all things related to ancient civilizations. By then, a distinction was being drawn between jewelery for evening wear and more restrained jewelery for daytime. Carnelian provided a bright pop of color without being inappropriately flashy. Carnelian cameos featuring scenes from Greek and Roman myths or the profiles of emperors and other illustrious figures became popular.

In the early years of the 19th century, engraved gems were popular. An exquisite neoclassical parure owned by Empress Joséphine, now in London's Victoria and Albert Museum, consists of a diadem, hair comb, earrings, and belt slide decorated with periwinkle-blue enamel and set with carnelian intaglios featuring the heads of Roman emperors and mythological scenes. The intaglios date from 100 BCE–200 CE.

Carnelian is sometimes mistaken for sard, and the names are often used interchangeably to describe chalcedony that is brown, red, or orange in color. In fact, sard is darker and browner than carnelian, ranging from dark reddish brown to black, while carnelian is a lighter reddish brown to orange.

For those who believe in the talismanic properties of stones, carnelian is purported to have many attributes. It is widely believed to be empowering, emboldening soldiers on the battlefield by giving them extra physical strength. In Arab and Muslim societies, it is considered to be a stone of kings, enabling them to speak eloquently and preserving their dignity. It is also believed to ward off envy and the evil eye, and is fashioned into protective "hands," sometimes engraved with prayers to reinforce its efficacy, just as

Signet rings

In the ancient world, carnelian was often incorporated into signet rings—finger rings carved with marks that identified their owner. Stamped into wet clay or hot wax, signet rings were used to authenticate important records and documents. They were also used as expressions of affiliation. In the early 20th century, German psychiatrist Sigmund Freud—whose own signet ring was a 1st century CE carnelian intaglio of Minerva, goddess of wisdom and justice, watching Victory crowning an enthroned Jupiter—gave signet rings set with ancient intaglios to his closest students and disciples. Known as the Secret Committee, this intimate circle of supporters promoted Freud's brand of psychoanalysis over that of Carl Jung.

> " All jasper and carnelian, he was bedazzled by the grandeur and opulence. "

Richard Francis Burton, *One Thousand and One Nights*, 1888

This portrait of the Empress Josephine painted by Andrea Appiani in around 1808 is believed to show her wearing a carnelian parure.

carnelian scarabs and bulls' heads were in ancient Egyptian times. In modern crystal lore, carnelian's fiery colors associate the stone with sexual attractiveness and passion. It is also believed to draw good fortune, and is sometimes worn by people hoping to succeed in business ventures.

Agate and gold brooch,
Italy c. 1860

Agate

❦

Dazzling patterns

One of the most remarkable examples of ancient art is the Pylos Combat Agate, a seal stone dating from 1450 BCE that was found in a tomb near the site of Pylos in Greece in 2015. The almond-shaped agate stone, just 1½ in (3.6 cm) in length, is intricately engraved with the scene of a warrior slaying two opponents. Archaeologists do not know how such a detailed and expressive image could have been carved without the help of a magnifying glass, but evidence of such a device has never been found in ancient Greece.

Agate is a banded gemstone that has many different, recognizable patterns. These have a host of evocative names, such as moss agate (translucent with dark green, red, and yellow inclusions), iris agate (which displays a rainbow when light shines through it), and fortification agate (with angular bands). Prized for its patterning, agate is carved into many different items that show these to best advantage, from cabochons, beads, and scent bottles to vases, bowls, and figurines. Because agate is porous, its colors are sometimes artificially enhanced by dye.

> **"** She is the fairies'
> midwife, and she comes
> In shape no bigger
> than an agate-stone. **"**

William Shakespeare, *Romeo and Juliet*, 1597

Throughout history and across cultures, agate has been seen as a protective stone. In the ancient Greek poem *Orphic Lithica*, the narrator, Orpheus, describes the magical properties of certain stones, including blue lace agate, which, he says, promotes persuasive speech, and moss agate, which he recommends tying to the horns of plowing oxen in order to guarantee a good harvest.

Many of the finest decorative items made from agate have come from Germany. In the 18th century, master craftsman and court jeweler Johann Christian Neuber used different agates sourced from across Germany to create some of the finest and most striking snuff boxes in the history of decorative arts. Each box was constructed using the *Zellenmosaik* technique of setting cut and polished pieces of hard stone into gold cage work.

Today, the German tradition of carving colored hard stones is synonymous with the Dreher family in Idar-Oberstein, who have been utilizing the patterns in agate for generations, and once produced pieces for Carl Fabergé. Dreher's carved agate birds and animals sometimes sell for tens of thousands of dollars.

Agate cup, Spain, 17th century

Jasper

❖ ━━━━━━ ❖

Guiding light

An opaque variety of quartz, jasper is cryptocrystalline, meaning that its crystals cannot be seen with the naked eye. It contains large amounts of impurities (up to 20 percent), which are responsible for its strong, earthy colors, ranging from red, brown, yellow, and orange to a deep green. These colors can be solid, which is relatively rare, or display a multitude of patterns, such as brecciation (fragmentation), possibly caused by earthquakes during its formation, banding, spots, and dendrites (branchlike structures).

The names given to the different types of jasper reflect their distinctive patterns; they include orbicular jasper (containing "orb-like" inclusions), leopard jasper (spotted), and landscape jasper (resembling rolling sand dunes). The origin of the name "jasper" is uncertain, but it may derive from the Hebrew word *yashepheh*, meaning "to polish," perhaps because jasper can be cut and polished to a brilliant shine.

Widely available in many parts of the world, jasper has been used as a talisman and for adornment throughout history. Some 12,000 years ago, Indigenous people in what is now Pennsylvania mined jasper to make arrowheads, knives, scrapers, and jewelery, and traded it with other peoples. In ancient Egypt, symbols and inscriptions from the Book of the Dead were carved into jasper amulets and buried with mummies to help guide their spirits in the afterlife. On the island of Crete, and elsewhere in the eastern Mediterranean, the Minoans (3500–1100 BCE) left thousands of jasper

Roman jasper and pearl clasp engraved with a hunting scene

> " So fair a church as
> this had Venus none:
> The walls were of
> discolour'd jasper-stone. "

Christopher Marlowe,
Hero and Leander, 1598

Green jasper necklace
with claw-shaped beads,
Japan, 7th century

seals inscribed with hieroglyphs and mythological figures. Often
set into rings, they were used to identify objects and documents.
In Norse legend, the hilt of Sigurd the Dragon Slayer's dagger was
made of jasper.

Jasper is good for carving, and has been used for functional
and decorative objects, from simple crochet hooks to rich inlay
in tables and cabinets. Often available locally, the stone was
frequently used in churches. In the Czech Republic, for example,

Bohemian jasper covers the walls and ceilings of St. Vitus, Prague's medieval cathedral, as well as Karlštejn Castle, northwest of the city. Mined in the nearby Ore Mountains on the orders of Emperor Charles IV (r. 1355–1378), jasper was also carved into small devotional objects, such as cups, rosaries, crosses, and reliquaries. Similarly, in the 17th and 18th centuries, jasper was used to decorate Europe's Baroque churches, its light and dark colors providing highly desirable contrast. Some 300 different types of jasper adorn Sicily's churches of the period. In the 20th century, the stone's bold colors made it an ideal choice for jewelery incorporating geometric designs.

In crystal healing, jasper is considered to be a tranquil and grounding stone, bringing calm and mental clarity. Certain jaspers are believed to have particular powers. For example, Dalmatian jasper, which is white or gray with black spots, is thought to promote playfulness, while brecciated jasper was traditionally believed to protect against spider bites. Rainbow jasper (featuring bands of dark red and gold) is said to prevent procrastination.

Jasper in the Bible

Jasper's ancient spiritual significance is acknowledged in the Bible. It first appears in the Book of Exodus (28:20) as one of the 12 stones embedded in the sacred breastplate of the Jewish High Priest, representing the 12 tribes of Israel. In the New Testament, John the Divine mentions jasper several times in the Book of Revelation as the first of the 12 stones used in the walls and foundations of the New Jerusalem, where it is said devout Christians will one day live with God. St. Jerome (c. 347–420 CE), an early translator of the Bible, and the monk Berengaudus (c. 840–892), an interpreter of the Book of Revelation, maintained that green jasper represented eternal life.

Jasper cameo in a gold frame studded with gems, Russia, 11th century

" ... and her light was like unto
a stone most precious, even like
a jasper stone, clear as crystal. "

The Bible, Book of Revelation, 21:11

Bloodstone

◇

The cure-all

Bloodstone, a variety of jasper, is mentioned in the records of several ancient cultures, including Sumerian and Egyptian, often in connection with mystical powers. In ancient Greece, it was known as heliotrope, derived from the Greek words *helios* and *trepein* ("sun turner") and was carved with images of the sun god Helios. It was said to be able to control the weather and warn of impending danger.

Dark green and dotted with bright red flecks or larger spots, the stone has long been associated with blood. According to one Christian story, the first bloodstone was formed at the crucifixion of Christ when drops of Christ's blood fell on a piece of green jasper at the foot of the cross. For this reason, it is sometimes known as Christ's stone. Its association with blood gave it supposed medicinal powers. In 15th-century Florence, Lorenzo de' Medici's physician prescribed bloodstone as a treatment for gout. Another variety of chalcedony that is dark green with yellow and white spots is called plasma.

Found worldwide, bloodstone is a cryptocrystalline quartz (its crystals are invisible to the naked eye). It can be carved into beads and intaglios, as well as larger items. In medieval Europe, it was often fashioned into talismanic rings, the protective powers of which were thought to be enhanced when messages were inscribed on the stone itself, or in the bezel of the ring.

In the 19th century, bloodstone was often carved with initials, signets, or coats-of-arms and used to imprint wax seals on

Bloodstone pendant set with a gold cameo of Emperor Charles V, France, late 19th century

" Whoever bears this stone ...
and pronounces the name
engraved upon it, will find all
doors open while bonds and stone
walls will be rent asunder. "

The Leyden Papyrus, c. 250 CE

letters. It was also fashioned into snuff boxes and bottles, and set into items such as cuff links. Some Cartier watches include bloodstone, and it has been used as a watch face for the Rolex Day-Date watch series. In modern jewelery, bloodstone is mainly used as beads or carved into cabochons and set into pendants, bracelets, and rings.

Chrysoprase

Green gold

Gold and chrysoprase choker plaque, c. 1900

Taking its name from *chrysoprasos*, ancient Greek for "golden green," chrysoprase is a variety of chalcedony colored green by nickel. Prolonged exposure to sunlight or heat can cause the color to fade. It is often fashioned as cabochons and set in rings and necklaces, or carved into beads and other small objects.

Chrysoprase became a popular gemstone in Europe in the 18th century, when large deposits were found in Poland. It was a favorite gemstone of Frederick II (the Great) of Prussia (r. 1740–1786). A chrysoprase snuff box commissioned by Frederick and given to his brother, now in London's Victoria and Albert Museum, is an exquisite example of Rococo decorative art. Carved from one piece of chrysoprase, it is decorated on all sides (including the base) with scrolling designs of leaves and flowers, all fashioned in varicolored gold mounts set with diamonds. Some of the diamonds are backed with colored foils to tint them pink, yellow, or green—the effect produced a beautiful miniature forest that could be kept in a pocket. Frederick II had several snuff boxes made from chrysoprase, and pieces of the gemstone were set out with his large collection of snuff boxes for him to enjoy during the last days of his life.

In the 1920s, chrysoprase was sometimes used to represent foliage in *giardinetto* ("small garden") brooches and earrings, usually in the form of flower baskets or bouquets, with colored gemstones representing the flowers.

Those who believe in the mystical power of gemstones view chrysoprase as a symbol for new life and personal and professional growth. It is also said to represent happiness.

“ Chalcedony drives away
phantoms and visions of the night. **”**

Joseph Gonelli, 1702

The Great Cameo of France, a five-layered
sardonyx cameo, 1st century CE

Sardonyx

The portrait stone

A variety of chalcedony, sardonyx is recognizable for its alternating white and brown bands in varying widths and transparencies. Hard stone carvers usually fashion sardonyx in ways that show off its variegated pattern, enhancing the stone's three-dimensional effect. It is well suited to cameos—gems carved in relief.

Sardonyx has been used for adornment for centuries. The ancient Romans were especially fond of the stone, which they carved into detailed depictions of emperors, gods, mythological scenes,

> ❝ In sardonyx, the humility of the saints, in spite of their virtues. ❞

Rabanus Maurus, Archbishop of Mainz, 9th century

and animals. High-ranking Roman soldiers wore sardonyx engraved with the image of Hercules or Mars, the god of war, to give them strength or courage. The Great Cameo of France, at 12 in x 10½ in (31 cm x 26.5 cm), the largest Roman cameo to have survived to the modern day, and now displayed in the Bibliothèque Nationale de France, in Paris, includes 24 figures representing emperors and mythological figures. It is thought to have been made to assert the legitimacy of the Julio-Claudian dynasty, founders of the Roman Empire (27 BCE–476 CE).

Sardonyx was used in a similar way from the 16th century, to depict portraits of Europe's kings and queens. The natural variation in the color bands from dark to light was often used to enhance the three-dimensional effect or to emphasize particular features. The Barbor Jewel, a sardonyx cameo of Queen Elizabeth I of England (r. 1558–1603), now in the Victoria and Albert Museum, London, uses the shading in the stone to highlight the queen's hair and ruff. According to tradition, the cameo was commissioned by William Barbor, a Protestant who escaped execution when Elizabeth ascended to the crown, replacing her Catholic sister Mary.

A passion for cameos was reignited in the 18th century, when sardonyx cameos referencing classical Greece and Rome were sometimes framed by precious gemstones.

Crystal healers assert that sardonyx enhances intelligence and improves communication. It is also said to promote the love of friends and help preserve long-term relationships.

Roman sardonyx carving in the form of a bear

SARDONYX

Blue Chalcedony

Glacial beauty

Some varieties of chalcedony have been individually named, such as carnelian or chrysoprase, but others are simply referred to by their color. Blue chalcedony is typically bluish-white to gray, but it can be a pale blue or mauve, and ranges from translucent to opaque with a waxy luster. Like all chalcedonies, it is cryptocrystalline (its crystals cannot be seen with the naked eye), yet it can sometimes grow in large formations that can be carved into many different shapes and forms.

Today, the best sources of blue chalcedony are Namibia, eastern Turkey, and the state of Montana, but it is found worldwide. Three-thousand-year-old Minoan seals made of blue chalcedony have been discovered in the eastern Mediterranean, and blue chalcedony cylinder seals for rolling on wet clay, dating from the 5th century BCE, have been excavated in Central Asia. The stone has long been made into jewelry designed to have protective powers, perhaps because of its mesmerizing inner luster, the result of light rays being scattered by layers of microscopic material within the stone.

Blue chalcedony was frequently used for gem engraving, both in ancient times and during the 18th-century European revival of Classical art and architecture. But it has also been embraced by modern jewelery designers. It was one of the favorite stones of Parisian jeweler Suzanne Belperron (1900–1983), who designed pieces for many of the great jewelery wearers of the day, including French writer and

Bull's head in blue chalcedony, Sumeria, c. 3000–2000 BCE

" A shimmering, floating, interior light. "

The Gem Society

socialite Daisy Fellowes, New York magazine editor Diana Vreeland, and American socialite Countess Mona von Bismarck. In around 1937, Belperron made a stunning blue chalcedony, diamond, and sapphire suite for the Duchess of Windsor, consisting of two cuff bracelets and a double strand of luminous blue beads with a large clasp in the shape of a flower. Today's designers have given the gemstone a new twist with statement pieces made from blue chalcedony that is botryoidal—possessing a bumpy surface reminiscent of a bunch of grapes, a formation that occurs in many chalcedonies.

In crystal healing, blue chalcedony is considered to be a calming stone, promoting the knowledge of inner truths. It is related to the throat chakra, which is purported to balance and release the intentions of all other chakras. Also associated with communication and speech, it is said to encourage reflection, holding back words that one may later regret. Actors are recommended to rub the stone on their throats and lips to help them overcome stage fright, and singers to soak the stone in water for an hour and then drink the liquid. This, it is said, will enable them to give voice to the stone's power.

Pendant with head of Medusa, blue chalcedony, France, c. 1885

Other Semiprecious Gems

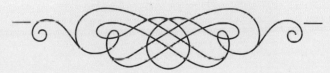

Topaz

The sun stone

Silver brooch set with topaz (orange), garnet (red), and emerald (green), early 18th century, France

Some of the world's largest cut gemstones are topazes, which occur naturally as huge crystals. In 2020, Guinness World Records officially verified one giant specimen as the largest cut topaz on Earth. The stone, belonging to Californian collector Dion Tulk, weighs 59,500-carats—nearly 26 lb (12 kg). Another showstopper was the so-called Braganza Diamond, which once formed part of the crown jewels of Portugal's royal family. It was found in the late 18th century in southeastern Brazil, then a

" The flaming topaz with its golden beam ... "

Richard Glover, *Leonidas*, 1737

Portuguese colony, and sent back to Lisbon, where it became part of the royal collection. The stone is said to have weighed around 1,700 carats. Although it was thought to be a diamond, it was almost certainly a yellow or colorless topaz. It later disappeared from the records, possibly following the royal family's exile in Brazil when Napoleonic French forces occupied Portugal in the early 19th century.

As a mineral, topaz is relatively abundant. It is an aluminum silicate containing the element fluorine. Present in fractures and cavities within rock that is cooling and solidifying while being pushed up toward the Earth's surface by volcanic action, bubbles of fluorine gas enrich molten lava and other hot fluids. As the fluids start to cool and solidify, so, too, does the fluorine. Given the presence of other constituent elements—aluminum and silicon—the conditions are right for large crystals of topaz to form.

In ancient Egypt, topaz was linked to the sun god Ra. Yellow topazes were also traditionally associated with wealth—perhaps because of their golden color—and thought to attract riches. The stone was also associated with extraordinary strength. In ancient Rome, topaz was believed to enhance a wrestler's might in line with the phases of the moon. The stone's power, it was said, increased when the moon waxed, especially if the full moon occurred in the astrological sign Scorpio. Another belief was that topaz amulets protected their owners from drowning, making them popular with sea captains

and others embarking on ocean voyages. Nowadays, crystal healers link yellow and fiery pink topazes with the sun's energy and endow them with restorative properties. Blue topaz, by contrast, is said to resonate with the blue vastness of the sky, evoking universal truth and compassion.

Valued for its sparkling brilliance, topaz is colorless in its pure form, but often contains traces of other elements that give it a range of colors, from sherry-brown to orange, yellow, blue, and pink. Occasionally, multicolored stones are found. Topaz mines are widely distributed around the globe, including in Myanmar (Burma), Pakistan, Sri Lanka, Zimbabwe, Namibia, and Nigeria, with Brazil the world's largest producer.

A golden topaz forms the center of this 19th-century papal clasp.

Connoisseurs prize the intense, saturated sherry yellows and musty pinks of the finest quality "imperial" topazes, originally mined in Russia. How these stones got their imperial designation may be due either to Russia's czars or Brazil's emperors. One explanation is that the Russian czars owned the mines in the Ural mountains where the cherished stones were found, and kept back the finest specimens for themselves. Another explanation is that in 1881—

> " An Oriental topaz, which may be pronounced unparalleled, exhibiting a luster like the sun. "

François Bernier, 1670–1671

when Brazil was still part of the empire ruled by a branch of the Portuguese royal family—the much-loved Emperor Pedro II was presented with a red-hued topaz when visiting the Brazilian mining town of Ouro Preto: from then on, such stones were known as imperial topazes.

The largest cut yellow topaz is the American Golden Topaz in the National Museum of Natural History, in Washington, D.C. From Brazil's Minas Gerais, an important mining state, this naturally yellow stone, cut and polished over the course of two years with 172 facets, weighs 22,892.5 carats—more than 10 lb (4.5 kg).

In 1750, a Parisian jeweler by the name of Dumelle made an interesting discovery: when he exposed a yellow Brazilian topaz to a high heat, it turned pink and stayed that color when the heat was removed. Then, in the 1960s, it was found that irradiation, followed by exposure to heat, turns a colorless stone, which is worth very little and bought by the ton, a deep blue. Because this can cause stones to become mildly radioactive, they are stored securely until the radioactivity has decayed to a safe

Seeing is believing

In the Rhineland of 12th-century Germany, abbess Hildegard of Bingen recorded the supposed healing power of topaz. A philosopher and composer of music, and venerated after her death as a saint, Hildegard was also a medical writer. Her *Causae et Curae (Causes and Cures)* describes remedies for various ailments, including "dimness" of the eyes. For this affliction, Hildegard recommends leaving a topaz in wine for three days and three nights. When the sufferer retired to bed at night, they were to rub their eyes with the moist topaz, allowing the moisture to spread over their eyeballs. Hildegard listed an additional benefit of the treatment: once the topaz had been removed at the end of the third night, the wine could still be drunk for up to five days.

level. Another common enhancement is to coat stones with metallic oxide to produce so-called mystic topazes, which change color in different lights.

Today, topazes are routinely "enhanced" in different ways, and it is not always possible to tell the difference between topazes that have been artificially colored and naturally occurring colored stones. Perhaps the most famous example of a "blue" topaz is the Ostro Topaz in London's Natural History Museum. Weighing 9,381 carats—roughly 4½ lb (2 kg)—and 6 in (15 cm) long and 4 in (10.5 cm) wide, the stone was found in Brazil in the mid-1980s by the British gemstone hunter and entrepreneur Max Ostro. He irradiated the original rough material to produce the most intense blue color.

The ability to produce vibrantly colored stones artificially has made topaz one of the most affordable stones, and even historic examples of naturally colored topaz have fetched only tens of thousands of dollars. In 2020, a yellow topaz parure that once belonged to Princess Amelia (1783–1810), the youngest daughter of King George III and Queen Charlotte, sold for £27,500 at auction in London. Another historic parure is the stunning Russian pink topaz demi-parure (it does not include a tiara) belonging to Queen Silvia of Sweden. Originally commissioned by Maria Feodorovna (1759–1828), Empress consort of Russia, the parure eventually passed into the hands of the Swedish royal family through marriage. Queen Silvia frequently wears the pink topaz parure on state visits and other important occasions, including the marriage of Victoria, Crown Princess of Sweden, in 2010. Like most parures, it is versatile. It can be worn in its entirety or split into smaller components. Two of the pendants on the corsage, for example, can be detached and worn as drop earrings.

A blue topaz cut into a heart shape

Queen Silvia of Sweden wears the Russian pink topaz parure, 2010

" Myriads of topaz-lights,
and jacinth work
Of subtlest jewelery. "

Alfred Lord Tennyson, *Morte d'Arthur*, 1485

Labradorite

Northern light

The iridescent sheen of
polished labradorite

R evered by the First Nations people of Canada for thousands of
years before Christian missionaries in Labrador, Canada, named
the stone in 1770, labradorite has a rich mythology. The Beothuk of
Newfoundland used the blue, green, and gold stone for healing and
decoration, believing it to be the frozen fire of the Aurora Borealis.
The Inuit, First Nations people of the Arctic, tell of a warrior who
discovered a cave filled with the light of labradorite. Believing this to
be the light of trapped stars, he slammed his spear down into the rock
to release them. Although the Aurora Borealis escaped, the remainder
stayed in the rock. It is this starlike quality that geologists have named
labradorescence, a measure of the stone's bright iridescence.

Labradorite is a type of feldspar, a crystalline mineral that
makes up over 50 percent of the Earth's crust. The vivid flash of
colors present in labradorite is not commonly seen in most

feldspars—it is a special inheritance, known as the Schiller effect, caused by light reflecting off different surfaces within the stone's structure. Although associated with Canada, labradorite is also found in Scandinavia and Russia. In Finland, it is associated with the mythical Bifröst Bridge, a shimmering rainbow bridge connecting the nine realms of Nordic mythology.

Because of labradorite's vivid colors and strong lustrous qualities, some practitioners of crystal therapy believe it can have a transformational effect on a person's career and personal life, reducing stress while increasing energy. In addition, the stone is associated with the third eye chakra (located between the eyebrows), considered to be the center of perception. It is believed that labradorite can help change negative emotions to positive ones, and enable one to remember dreams.

A fashionable gem, found in many modern jewelery ranges, labradorite looks best when smoothed and polished in a machine known as a rock tumbler. But labradorite is also an architectural stone, quarried in large pieces and cut into slabs for sculptures and table tops.

> " The colors of the male peacock—iridescent blue-green and amber gold await. "
>
> **Andrew Kerr-Jarrett**, 2022

Lapis Lazuli

The most perfect color

Lapis lazuli outlines the eyes and eyebrows on a bracelet found on the body of the pharaoh Shoshenq II, c. 885 BCE.

An intense and unrivaled shade of blue, lapis lazuli has been prized by civilizations for at least 6,500 years. The gemstone was transported from mines in what is now Afghanistan to Egypt, India, Mesopotamia, and beyond, and can be found in some of the most important—and beautiful—artifacts of the ancient world. In ancient Egypt, it was commonly carved into scarabs, pendants, and beads but was also used for more intricate work, such as cloisonné, in which filaments of gold or other metals frame and hold in place cut

> " Hail, dweller in the lapis-lazuli!
> Watch ye over him that
> is in his cradle, the Babe when
> he cometh forth to you. "

The Book of the Dead, Papyrus of Ani, 1250 BCE

gemstones and colored materials. An exquisite necklace and pectoral (chest pendant) from the tomb of a 19th-century BCE Egyptian princess, Sithathoryunet, and now in the Metropolitan Museum of Art in New York, consists of 372 lapis lazuli, carnelian, turquoise, and red garnet stones set in gold.

For centuries, artists used lapis lazuli to make a blue pigment. Sourced from the mountains of northeast Afghanistan, it was used as a pigment in Buddhist temples in the 6th century CE. A soft gemstone that gets its color from the mineral lazulite, it can easily be ground down to make ultramarine, once described by Italian artist Cennino Cennini (c. 1370–c. 1440) as "the most perfect of all colors." Cennini wrote instructions on how to prepare this stunning blue pigment.

Used by European artists from the 13th century, when lapis lazuli was probably introduced to Europe, ultramarine is common in Renaissance art. Titian's *Bacchus and Ariadne* (1522–1523), for example, contains ultramarine pigments in its brilliant blue sky. Many artists used it to paint the robes of the Virgin Mary, the color blue having been associated with the "Queen of Heaven" since the 5th century. Its cost—equivalent to the price of gold—was considered fitting for the Virgin

Gold and lapis lazuli
earring, Egypt,
c. 1295–1186 BCE

and enhanced the status of paintings when it was used. In the 17th century, Dutch artist Johannes Vermeer used ultramarine in almost all of his paintings, sometimes to the exclusion of all other blue pigments. Ultramarine derived from lapis lazuli became less popular after a synthetic and less expensive alternative was created in 1826.

In the Russian Empire, lapis lazuli was prized as a symbol of the country's wealth and power. Czar Nicholas I (r. 1825–1855) commissioned giant urns faced with lapis lazuli for his Winter Palace in St. Petersburg. The stone was transported more than 1,240 miles (2,000 km) from stone-cutting factories in Kolyvan, in Central Asia, to Yekaterinburg, in the Urals, where it was affixed to the 6-ft- (2-m-) tall marble urns. In 1912, Nicholas II commissioned Fabergé to make a lapis lazuli imperial egg for his wife, Alexandra Feodorovna.

Lapis lazuli was once believed to protect the wearer from malevolent spirits, and was often fashioned into evil eye jewelery or made into amulets. Its reputation as a protective gem still applies today. In crystal healing, it is said to ward off negative energy and psychic disturbance. It is also associated with the throat and third eye (forehead) chakras, and believed to promote communication and intuition.

Elvis Presley's rings

Rock legend Elvis Presley (1935–1977) adored flashy jewelery of all kinds, but particularly rings. Among these were several lapis lazuli rings, lapis being one of his favorite stones according to his best friend Charlie Hodge. Presley left the bulk of his jewelery collection to his manager, Colonel Tom Parker, but he regularly presented rings to friends and associates during his lifetime, and examples sometimes come up for auction. The ring shown left, a lapis oval flanked by 14 diamonds mounted in 18 carat gold, which Elvis is known to have worn on and off stage on several occasions, sold for $33,750 in 2009.

" A noble color, beautiful, the most
perfect of all colors. "

Cennino Cennini, *The Craftsman's Handbook*, c. 1437

Amazonite

For fluid thoughts

Known as the "Stone of Hope," amazonite has had a long and mystical history. Amazonite from East Africa has been found in Neolithic cemeteries in the Sudanese Nile Valley, indicating trade in the stone from around the 6th millennia BCE. By the time of the ancient Egyptians in around 3100 BCE, it was being used with other semiprecious blue-green stones in sophisticated jewelery and amulets to ward off evil.

The stone takes its name from the Amazon River—its hazy green hues are the result of lead and water in its potassium feldspar structure. Although no deposits have been discovered in the river itself, amazonite is found elsewhere in Brazil, as well as in Colorado and Virginia, in Australia and Madagascar, and in smaller quantities elsewhere around the world.

In crystal healing, amazonite is associated with purification, stillness, and calm, and is said to be beneficial in times of stress. Crystal healers also claim amazonite is intrinsically linked to the heart and throat chakras. It is believed to help rebalance the chakras, allowing for more fluid thoughts and making chaotic ones float away. By opening the throat chakra, amazonite is said to aid in communication, helping people to stay true to themselves and to set boundaries, both personally and with others. Because of its relationship to the heart and throat chakras, amazonite is most often worn on a necklace, to bring the wearer tranquility and balance.

Amazonite in its rough state

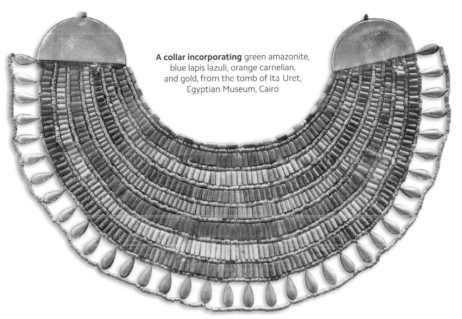

A collar incorporating green amazonite, blue lapis lazuli, orange carnelian, and gold, from the tomb of Ita Uret, Egyptian Museum, Cairo

" Blue-green semiprecious stones were used to imbue objects with symbolic power. "

The Metropolitan Museum of Art

19th-century moonstone and ruby ring, Cambodia

Moonstone

◇

Heaven sent

With an inner glow that resembles moonlight, this gemstone has been associated with the moon since antiquity. Ancient Greeks named it *selenites* after their moon goddess Selene, and in ancient India it was called *chandrakanta*, Chandra being the Hindu moon god and *kanta* meaning light. The Romans, who thought the stone was solidified moonlight, believed it to be a gift from the moon goddess Diana. They carried it with them to bring luck and foresight.

Moonstone is milky white or colorless. The best examples, which are colorless but have an intense blue shimmer, come from Sri Lanka. The shimmer, an effect known as adularescence, is caused by light diffracting through the internal layers of the stone; this moves as the stone is turned.

In some cultures, gems such as moonstone that appear to be lit from within are associated with good luck, as the moving light indicates that a good spirit resides in the stone. Because the moonstone is linked to moonlight and the night, it is also

> ❝ He loved the red gold of the sunstone,
> and the moonstone's pearly whiteness. ❞

Oscar Wilde, *The Picture of Dorian Gray,* 1891

associated with love, dreams, and tranquility. And, owing to the moon's influence over oceans and tides, it was once thought to provide protection during sea journeys.

Moonstones are most often cut en cabochon (round or oval in shape with a smooth dome) to show off their inner glow. They became popular in Art Nouveau jewelry in the early 20th century. French designer René Lalique used moonstones extensively in his creations, as did the American designer Louis Comfort Tiffany, the son of Tiffany's founder Charles. Originally a stained-glass designer, Louis loved the inner sheen of moonstone and often paired it with sapphires.

Red Garnet

Inner fire

Deep red with a fiery inner glow, red garnets were especially popular in Victorian times. The name "garnet" comes from the Latin *granatus*, meaning "seed" or "grain," probably because the stone is reminiscent of the seeds of the pomegranate fruit. In classic 19th-century garnet jewelery, small rose-cut stones were often set in clusters. Sometimes, the small stones surround contrasting larger ones, often cut en cabochon.

Garnets are a group of crystalline minerals that have similar structures but different chemical compositions. At least 15 distinct types of garnet exist, but of these only five are widely valued as gemstones. The principal red garnet varieties are

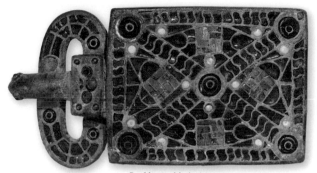

Buckle studded with red
garnets, Spain, c 6th century

" There shall it flame for ages,
making a noon day of midnight. "

Nathaniel Hawthorne, *The Great Carbuncle*, 1835

almandines (containing iron and aluminum in combination with
silicon and oxygen) and pyropes (containing magnesium and
aluminum with silicon and oxygen). Red garnet crystals are very
rarely found in their pure form. Most of them contain both iron
and magnesium atoms, with a preponderance of one or the
other. Pyrope-rich stones are highly prized for the dark intensity
of their color and the fieriness of their inner glow. The name
"pyrope" comes from *pyropus*, an ancient Greek word meaning
"fire-like" or "fire-eyed."

Since earliest times, humans have cherished red garnets
for their beauty and attributed magical properties to the stones.
Inevitably, their color associated the stones with blood and the
heart, while their glow connected them with spiritual energy or
life force. Garnets were believed to stimulate the circulation, and
to "give heart"—or embolden—and were associated with military

Bracelets incorporating large red garnets and other gems, Greece, 1st century BCE

<blockquote>
" Beautiful crimson lights flashed suddenly, deep under the smooth egg-shaped surface of the stones. "
</blockquote>

Alexander Kuprin, *The Garnet Bracelet*, 1911

prowess. In the mountains of Kashmir, warriors in the high valley of the Hunza fired red garnet pellets, originally using bows, later from guns as bullets. For them, the stones' blood color suggested deadliness. British colonial forces attacking the Hunza region in the 1880s and '90s found themselves under fire from these blood-red missiles.

Red garnets are associated with spiritual as well as physical strength. Ancient Egyptian tombs have yielded several red garnet jewels. In one burial dating from around 3500 BCE, archaeologists found the body of a young man wearing a necklace with beads and fragments of carnelian, shell, turquoise, amethyst, and almandine garnet. The Egyptians also used garnets in cloisonné work, in which filaments of gold or other metals frame and hold in place cut gemstones and colored materials.

Gold and garnet cloisonné was popular in Europe after the decline of the Roman Empire (from 476 CE). The Franks, a militaristic people who invaded the Western Roman Empire in the 5th century, used gold and garnet cloisonné to embellish dagger hilts, belt buckles, and the brooches that secured their cloaks. Around 300 gold and red garnet cloisonné insects were among treasure discovered in the tomb of Childeric I, a 5th-century CE Frankish king, which a Belgian builder stumbled upon in 1653. Along with the cloisonné insects, which probably represented bees, was a gold and red garnet sword hilt.

In the Middle Ages, red garnets were associated with dragons. It was said that dragons' eyes were made of glowing garnets. There was also a tradition that garnet amulets could warn travelers of danger. In North America, the Pima people, in what is now central and southern Arizona, believed that garnets were an antidote to poison.

European gem traders sourced most red garnets from Sri Lanka. In around 1500, Europe found an abundant source of high-quality, pyrope-rich garnets in the highlands of central Bohemia, north of Prague, in what is now the Czech Republic. For four centuries, this region was the world's garnet capital. At London's Great Exhibition in 1851, a display of garnets from mines belonging to the great Bohemian landowner Prince von Lobkowicz (1797–1868),

Mandarin garnet

Flame-bright orange stones—called mandarin garnets—are the most striking and highly prized specimens of spessartite, one of the five gem "species" of garnet. Containing manganese and aluminum combined with silicon and oxygen, spessartite ranges in color from orange to reddish brown. Examples of the most brilliant orange stones, hard enough to be cut, used to be extremely rare. But this changed in the 1990s when gem hunters found them in commercial quantities in Namibia in southern Africa. Similar stones, also categorized as mandarin garnet, were later discovered in Nigeria and Tanzania.

was among the star attractions. By the end of the century, the Bohemian mines were becoming less productive, but other good sources of red garnets were emerging, notably in South Africa where garnets accompany the diamond deposits around Kimberley in Cape Province. These stones were often (inaccurately) referred to as "Cape rubies."

Although not quite as prized as they once were, red garnets are still appreciated for their pomegranate luster. They feature in many royal jewelery collections. The lovely Rosenborg Garnet Kokoshnik, a pearl, diamond, and garnet tiara, was commissioned in the 1930s by Prince Viggo, Count of Rosenborg for his American-born wife (née Eleanor Margaret Green), a commoner for whom he gave up his place in the Danish line of succession. The tiara eventually passed to Countess Ruth of Rosenborg (1924–2010), who frequently wore it on royal occasions. In recent years, one of the highest-profile items of garnet jewelery was a Victorian garnet and pearl ring that Britain's Prince William gave his future wife, Catherine Middleton, while they were dating—garnet is the birthstone for January, Catherine's birth month, and pearl for June, the month William was born.

Carbuncles

Nowadays, the term "carbuncle" refers to an almandine garnet (the most common gem garnet species) cut en cabochon. Until the 19th century, however, it could refer to any cabochon-cut red gemstone, including ruby and spinel. Such stones gathered all kinds of associations. One story from southeast Asia told of a snake that slithered into a hammock slung between two trees where a baby prince lay sleeping. Instead of killing the prince, the snake dropped a rare carbuncle onto his sleeping body. Out of gratitude, the child's parents cared for the snake for the rest of its days.

Garnet and gold
brooch, England,
19th century

" If you would cherish friendship true,
In Aquarius well you'll do
To wear this gem of warmest hue—
The garnet. "

Saying quoted in George Kunz's *The Lore of Precious Stones*, 1913

Green Garnet

Hidden depths

A "trilliant-cut" green garnet

Demantoid, a green variety of garnet, is among the most prized of all garnet gemstones. It is a variety of the garnet type andradite, which is rich in calcium and iron, but its crystals obtain their lustrous grass- or olive-green color from traces of chromium. For dispersion—the capacity to break light down into the colors of the rainbow—demantoid outshines diamond.

Green garnet was first discovered in Russia's Ural Mountains in the early 19th century, making it a latecomer among gemstones. It rapidly won favor at the Russian Imperial court, and at the end of the century its color suited the palette of Art Nouveau jewelers, who used it widely in their designs. The Finnish mineralogist Nils Gustaf Nordenskiöld called the stone "demantoid" after an old German word for "diamond." Russia is still the principal source of the stone, although in recent decades commercially viable deposits have also been mined in Namibia and Madagascar. It is also found in small quantities in Italy, Afghanistan, and Iran. Other green garnets are uvarovite and tsavorite. The former was also first discovered in the Ural mountains, in 1832, by Swiss-Russian chemist Germain Henri Hess, who named it after the prominent scholar and amateur mineral collector Sergey Semyonovich, Count Uvarov. A chromium-rich species, uvarovite is among the rarest of all garnets, with crystals of a gorgeous emerald green. Unfortunately for jewelers, its crystals are mostly opaque with only tiny transparent portions suitable for facet-cutting.

Tsavorite, by contrast, is hard, transparent, and ideal for faceting. It is a variety of the garnet species grossular, rich in calcium and aluminum, with traces of chromium or vanadium, giving it its brilliant green hue. Another latecomer to the gem trade, it was discovered in Tanzania in 1967 by British geologist Campbell Bridges; he named it after Tsavo East National Park across the border in Kenya. The New York jewelers Tiffany & Co. promoted tsavorite extensively in the 1970s.

" A green color
that rivals emerald
and a fire that
exceeds diamonds. "

International Gem Society

Green garnet, diamond, and
ruby dragonfly brooch

Rhodonite

The rose stone

Pink or rose-red hues give rhodonite its appeal: the name comes from the Greek *rhodon*, meaning "rose." Chemically, it is a manganese silicate—meaning manganese combined with silicon and oxygen—found in or near manganese-bearing ore rocks. Most commonly, rhodonite takes the form of an opaque rose-red stone, strikingly shot through with black dendritic (treelike) veins of manganese oxide. The opaque crystals can be cut en cabochon, shaped and polished into beads, or carved into decorative objects, figurines, boxes, and bowls. Much rarer are transparent crystals of brilliant, almost shocking pink rhodonite. These make spectacular jewels when cut with flat facets, although the process is notoriously difficult, since the crystals fracture easily.

> " The rhodonite is typical massive
> pink material, delicately veined. "

John Sinkankas, *Gemstones of North America*, 1959

Rhodonite was first mined on a commercial scale in Russia's Ural Mountains at the end of the 18th century. Known as *orletz*, "eagle stone"—supposedly because eagles took pebbles of rhodonite to line their nests—this Russian variety of the stone was opaque with a particularly intense rose color. It was fashioned into figurines, clocks, boxes, and other ornaments at the Imperial Lapidary Works in Ekaterinburg, which helped to popularize the stone, especially in the late 19th century when Russia became a favored destination for wealthy European travelers. German naturalist Christoph Friedrich Jasche gave the stone its name "rhodonite" in 1819.

The Urals are still the biggest source of rhodonite. Other mines exist in British Columbia (Canada), New South Wales (Australia), Sweden, Japan, Tanzania, parts of the US, Mexico, Brazil, and Peru. Most of the stones that are suitable for jewelery come from Japan and Australia.

Rhodonite is also popular with advocates of crystal healing. The attractive pink stone is believed to help soothe and focus the spirit, encouraging forgiveness and the release of trauma. It is also said to nurture love while calming an over-active libido.

Kremlin clock tower by Fabergé, with rhodonite, silver, and emeralds, c. 1913

Jade pendant,
Liao dynasty
(916–1125 CE)

Jade

◇

A dragon's tears

Humans have been entranced by jade for millennia. To the Aztecs of Central America it was worth more than its weight in gold. But nowhere has valued jade more than China, whose people shaped it into intricately carved tubes and disks—the purpose of which is unknown—from around 6000 BCE. By the time of China's first dynasties, around the 3rd century BCE, it was intrinsically linked to imperial power. According to one legend, when China was invaded by the Mongols in the 13th century, the tears of dragons turned to jade as they fell to the ground.

Jade plays an important role in Chinese philosophy, in both Confucian and Daoist teachings. According to the *Shuowen Jiezi*, a dictionary written during the Han dynasty (206 BCE–220 CE), the stone was endowed with five virtues: "charity—embodied by its bright yet warm luster; purity—symbolized by its translucency; wisdom—expressed by the clear note that rings out when the

> " Hit a piece of quartz and it will split in two. But if you hit a piece of jade, it will ring like a bell. "

Jewelers' saying

stone is struck; courage—as the stone can be broken but not bent; and equity—represented by its sharp angles that injure no one." These five cardinal virtues were so important to Confucius that he advised men to wear jade pendants to guide their behavior. The chime of small pendants as they struck one another, suspended on a girdle or from headgear, reminded their wearers of these cardinal virtues.

The ancient Chinese also believed that jade had healing powers. Raw jade ground and dissolved in wine or water was used to relieve heartburn and asthma, and as a general tonic. It was believed to make the hair shine and to strengthen the voice. Under the Han dynasty, the ruling classes were buried in suits made entirely of jade pieces, woven together with thread—gold wire for emperors; silver for princes, princesses, dukes, and marquises; and silk thread for lesser nobility. Believed to stop demons from entering the body, the jade suits covered every inch of flesh, each piece sewn tightly to the next. They also helped prevent decay—a priceless protection for a people who were preoccupied with the afterlife. Jade objects were often placed in tombs and jade "spirit tablets" were used to record the names of ancestors.

Ritual masks

Jade had an important ritual purpose for the Olmec, people who lived on the Gulf of Mexico from 1000 to 400 BCE, as it did for later Mesoamerican people, such as the Mayans and Aztecs. Associated with water, growth, and life, jade was carved into miniature masks (see left), probably for ceremonial purposes. The masks were mostly human looking, but had exaggerated features evoking the Olmec gods. Many Olmec masks, for example, represent the maize god. When the Spanish conquistadors invaded Mesoamerica in the 16th century, they were surprised to find that jade was considered more valuable than gold.

Chinese artisans used jade to make jewelery, ornaments, ritual objects, and statues, and incorporated it into furniture. Jade carving flourished under the Qing dynasty's great Qianlong Emperor (r. 1735–1796), after he conquered jade-rich lands in Eastern Turkistan. The biggest ever jade sculpture, *Yu the Great Taming the Waters*, 7 ft 3 in (2.24 m) tall and 3 ft 1 in (96 cm) wide, was made during this period, taking a team of sculptors more than seven years to craft. Representing one of China's most beloved myths, of the legendary King Yu controlling a great deluge, it depicts Yu's workers laboring in a precipitous landscape of caves and forests.

Jades from the Qing period are in high demand today; one of Emperor Qianlong's imperial seals, a white jade decorated with intertwining dragons, sold at Sotheby's for $18.8 million in 2021, while a green seal of the emperor's sold for $15.7 million in 2010.

While jade is usually associated with the color green, it comes in many different hues, which are determined by the presence of trace elements. In China, these colors were believed to correspond to the elements. Blue jade was associated with the

White jade pendant depicting a monkey in a grapevine, China, 8th century

❝ Gold has a value; jade is invaluable. ❞

Chinese saying

sky or heavens; yellow with the earth; green with wood and the spring; red with fire; white with metal; and black with water. Over time, the availability of jade from different sources and changing fashions led to colors other than green becoming popular. Among the rarest are "mutton-fat" jade, a marbled white, and the vibrant green of "imperial jade," so called because the imperial court had first choice of any stones of that shade.

Despite jade's long history, it was not until 1863 that anybody realized that more than one stone was known as jade. That year, French mineralogist Alexis Damour studied two jades—one from Burma (now Myanmar), and one from China—and found that their chemical compositions were different. The two types of jade are now known as jadeite and nephrite. It can be difficult to tell the two stones apart with the naked eye, but below the surface they are very different. Jadeite is made up of grainy crystals, while nephrite looks fibrous, with interlocking crystals.

Jade's importance in China is well known, but the stone is found in many parts of the world, and other cultures have also celebrated its beauty and supposed

" Soft, smooth, and glossy, it appeared to them like benevolence. Fine, compact, and strong – like intelligence. "

Confucius, *Book of Rites*

18th-century nephrite jade turban pin set with rubies, emeralds, and rock crystal

curative powers. Nephrite's name comes from the Greek *nephros*, meaning "kidney," whereas jadeite is named for the Spanish *piedra de ijada*—"stone of the side," reflecting a belief among the Aztecs that holding a piece of jade against the side of the body could cure kidney or back pain. The nephrite form of jade was also important to the Māori of New Zealand, who carved amulets called *hei-tiki* from jade. According to one of many Māori legends about jade, the water being Poutini created jade from a woman called Waitaiki, whom he had abducted. Chased and eventually cornered by her husband Tama-ahua, Poutini turned Waitaiki into the precious stone.

In Mughal India (1526–1761), jade imported from Kashgar in Central Asia was fashioned into intricately decorated dagger hilts and luxurious objects. The wine cup of Emperor Shah Jahan (1592–1666), an exquisite white nephrite cup with a handle in the shape of a ram's head and a base in the form of lotus flower, is considered one of the finest surviving objects of the Mughal court. Jade was also used to hold the jewels in elaborately decorated turban ornaments, as an alternative to gold. As symbols of kinship, and passed from father to son, turban ornaments were vehicles for displaying the emperor's most impressive gems.

Spirit holder

The Māori, the indigenous people of New Zealand, call jade *pounamu*, the "God stone," because it is believed to bring life and protection. Jade *hei-tiki*, pendants carved into human shapes and worn by men and women, are often passed down from generation to generation within a family, because they are believed to hold the spirits of its ancestors. A *hei-tiki* is treasured because it is thought to have its own personality. These pendants can be seen around the necks of Māori people in some of the oldest depictions of their culture. Today, the pendants have become emblematic of both Māori people and New Zealand.

> " Jadeite stones are art creations gifted by Mother Nature. "

Wenhao Yu, Sotheby's Asia

Jadeite and nephrite are both found in Russia, but the country is particularly known for the dark, spinach-green nephrite mined in Siberia. Among the most beautiful examples of Russian "spinach jade" is the "Pansy Egg," one of 52 jeweled eggs created by the St. Petersburg House of Fabergé for the Russian imperial family. Commissioned in 1899 by Czar Nicholas II as an Easter gift for his mother, Maria Feodorovna, the Pansy Egg is a hollow nephrite egg exquisitely decorated with pink enamel flowers and diamonds; inside the egg were portraits of the imperial family.

The opening of the Musée Guimet, which specializes in Asian art, in Paris in 1889 popularized jade in France, and dealers in the gemstone set up offices in the city. Cartier sourced historic jades for Art Deco pieces in the early 20th century, taking inspiration

Hutton–Mdivani necklace

Comprising 27 thick jadeite beads, the Hutton-Mdivani necklace was a gift to Woolworth heiress Barbara Hutton from her father on her marriage to Georgian Prince Alexis Mdivani in 1933. Considered the greatest jadeite bead necklace in the world, it is now part of the Cartier Collection. Cartier acquired the necklace at auction in 2014, paying $27.44 million after an intense bidding war. It was among 26 jade treasures exhibited by Cartier at the Musée Guimet in Paris in 2017, the centerpiece of what was billed as "an eastern dream in the heart of a Paris winter."

from the arts of the East and adding Western elements. Some of these jade pieces were hundreds of years old—one pocket watch from 1929 incorporated a 14th-century Chinese jade carving of a Buddhist lion on a pedestal.

The allure of jade has not faded. While the Asian jade market prizes imperial jade, realizing sky-high prices at auction, designers catering to the Western market use the full jade color palette in creative ways. Hong Kong jeweler Michelle Ong's brand Carnet has made many standout jade pieces. One of its most colorful items, its "glowing jade and gems" necklace, consists of green, lavender, and yellow jade pieces combined with diamonds in a stunning pharaoh-style bib. Other exciting designers working in jade include German jeweler Hemmerle and American-born jeweler Joel Arthur Rosenthal (JAR), who is based in Paris.

Chrysoberyl
brooch, UK,
19th century

Chrysoberyl

Eye-catcher

Typically yellow-green in hue, chrysoberyl was popular in the 18th and early 19th centuries, especially in Portugal, which imported the gemstone from its colony in Brazil. Capable of being faceted, but less costly than diamonds, it was used for both jewelery and devotional objects, often backed by foil to emphasize the stone's sparkle.

Some chrysoberyl stones possess a striking optical effect called chatoyance, also known as "cat's-eye effect," caused by microscopic channels running through the crystals. When a stone is cut en cabochon (with a dome on top), light reflected through the stone creates a shining arc across the center of the dome, resembling the

> **“** Dull fires seem to
> glow within them. **”**

G. F. Herbert Smith,
Gemstones, 1940

elliptical shape of a cat's pupil. Cat's-eye stones also have a slight
cloudiness, and some give the impression of being much darker on one
side than on the other, an optical illusion known as "milk and honey."
While cat's eyes occur in other gemstones, including tourmaline and
apatite, they tend to be best known in chrysoberyl.

The world's largest cat's eye chrysoberyl is a 465-carat cabochon
known as the Eye of the Lion. Discovered in Sri Lanka in the late 19th
century, the gem, then uncut, belonged to Iddamalgoda Kumarihami, a
wealthy landowner. Cut and set into a necklace by her grandson in the
1930s, the gem reportedly caught the eye of the Maharaja of Baroda
(India), a man renowned for his love of fine gems.

Chrysoberyls are prized as charms to ward off evil spirits. Placed
next to the bed, the gem is said to keep nightmares at bay. Crystal
healers also claim that it can boost confidence and concentration.

Aquamarine

Sailor's treasure

A blue variety of beryl, aquamarine is common in Brazil, and many of the world's most famous examples have been given as gifts by the Brazilian state. An aquamarine parure belonging to the late Queen Elizabeth II includes an aquamarine and diamond necklace and earrings given to her by Brazilian president Getúlio Vargas on her coronation in 1953. She commissioned London jeweler Garrard to create a tiara to match these items, and a bracelet and brooch were later added, producing a suite of jewels that she wore on many occasions.

The name aquamarine is derived from the Latin words *aqua* (meaning water) and *mare* (meaning sea), and reflects the stone's sea-blue color. The ancient Greeks wore it as a lucky charm to calm the sea and protect them from drowning, and legend has it that it

The Dom Pedro aquamarine

brought good fortune to those who found it washed up on beaches. The stone's association with tranquility is still asserted by believers in the healing properties of gemstones.

Aquamarine's color is caused by traces of iron and ranges from pale to deep blue. Crystals can be large—often up to 3 ft (1 m) in length. The largest gem-quality crystal ever discovered was found in Marambaia, Brazil, in 1919. Measuring 18 in (48 cm) long and weighing 243 lb (110 kg), it was cut into a large number of gems totaling around 100,000 carats. The largest-known cut aquamarine, also from Brazil, is a 13½-in- (35-cm-) high carved obelisk known as the Dom Pedro, now in the collection of the Smithsonian Institution in Washington, D.C.

Depth of color is the most important quality when it comes to value. Stones should be a clear blue-green. Being large with few inclusions, aquamarines are easy to cut, and gem cutters often experiment with new shapes. Aquamarines are ideal for cocktail rings. A stunning aquamarine ring that once belonged to Diana, Princess of Wales, is now owned by Meghan Markle, Duchess of Sussex. Worn by Meghan on the evening of her wedding to Prince Harry in 2018, the ring is an emerald-cut aquamarine surrounded by small solitaire diamonds and mounted on a 24-carat gold band.

> **"** The aquamarine ... has the glassy tint of the waves of the sea. **"**
>
> **Charles Blanc**, *Art in Ornament and Dress*, 1875

Alexandrite

◆————◆————◆

Russian rarity

" The jewel ... is concentrated
brilliancy, the quintessence of light. "

Charles Blanc, 1875

Alexandrite is a rare and extraordinary variety of chrysoberyl
that was allegedly discovered in 1830 in the Ural Mountains
in Russia. It was named after the Russian crown prince, the future
Czar Alexander II, by mineralogist Count Lev Alekseyevich Perovski
when Alexander came of age a few years later. Its red and green
hues reflected Imperial Russia's military colors.

The gemstone was thought to be emerald until it was
found to have a secret, seemingly magical property: an ability to
change color. Sometimes described as an "emerald by day, ruby
by night," Russian alexandrite is a rich green color in daylight or
any light that has a higher concentration of blue wavelengths.
Change the source of lighting to a candle or an incandescent
light bulb, and the stone smolders exotically with gleams of
red. This quality made alexandrite a popular jewel to wear
at candlelit balls.

The best quality alexandrite displays
dramatic color change from green to
red, or, in the case of Brazilian
alexandrite, from bluish-green to
purplish-red. The more extreme
the color change and
depth of color,

Alexandrite ring, c 1960

combined with the absence of inclusions, the higher the value of the stone. The highest quality natural alexandrite can be as expensive as rubies and diamonds.

Antique Russian alexandrites are still valued for the intensity of their colors, but nowadays most alexandrite comes from Brazil, Tanzania, Mozambique, Madagascar, India, and Sri Lanka. An even rarer and more coveted variety of alexandrite is cat's eye alexandrite. This gem not only shows the color-change effect but also displays a bright band of light (like a cat's eye) that seemingly hovers over the surface of the gemstone when it is illuminated with a pinpoint of light.

Cut and polished
morganite

Morganite

◇

Rose-tinted sparkle

Shimmering pink morganite is a beryl, a member of the same family as emeralds and aquamarine. A find in Madagascar in 1910 first brought it to the attention of Western collectors. Until then, where known, it was simply referred to as rose or pink beryl. The influential gemologist, George Frederick Kunz, a director of the New York jeweler Tiffany and Co., proposed the name "morganite" to members of the New York Academy of Sciences in tribute to his friend and patron, the banker John Pierpont Morgan. Kunz had assembled two important gem collections for Morgan that were later donated to New York's American Museum of Natural History.

Morganite owes its color to traces of the metal manganese. While some stones are heat-treated to make their color more consistent, connoisseurs appreciate the untreated, natural stone. Classic morganites are rose pink, but hints of orange give some a peach color, while others are lilac.

Today, most morganite comes from Brazil, Madagascar, and California, though one of the largest crystals ever found came from a mine in Maine in 1989. Christened the Rose of Maine, the uncut stone weighed an impressive 115,000 carats (50 lb/23 kg).

Heliodor

Desert gold

In 1910, workers on the bare granite mountain of Klein Spitzkoppe in the deserts of western Namibia discovered beautiful golden-yellow beryl crystals. The name heliodor, a portmanteau of the ancient Greek words *helios* (sun) and *doron* (gift), was initially a trade name for the Klein Spitzkoppe gems, but the term was later broadened to include a range of beryls, some golden-yellow, like the first "gifts of the sun" from Namibia, others greenish-yellow. Namibia is still a major producer of heliodor, but it is also found in Madagascar, Ukraine, Russia, and Brazil.

Like other members of the beryl family, including morganite, emerald, and aquamarine, heliodor is an aluminum beryllium silicate. The presence of small amounts of iron and uranium in the crystals gives it its color; the uranium can also make it slightly radioactive (at non-dangerous levels). Like all beryls, heliodor is relatively hard, making it suitable for jewelery that has regular use, and it is an ideal stone for cocktail rings. It can be found in substantial crystals, big enough to be carved as ornaments. The world's largest cut heliodor, recognized by *Guinness World Records*, is a 5,900-carat (2.6 lb/1.18 kg) stone belonging to a Californian collector.

Practitioners of crystal healing assert that heliodor's golden luster evokes the life-giving energy of the sun, and link it to enlightenment, or higher knowledge. Wearing heliodor is said to bring hope and confidence.

Heliodor in its uncut state

Peridot

Star appeal

The yellow-green peridot is the transparent, gem-quality variety of the mineral olivine. The stones, which are sometimes of considerable size and usually step cut, were particularly fashionable in the mid- to late 19th century. Some of the most spectacular examples featured in the Hapsburg diamond and peridot parure, a bandeau-style tiara, necklace, earrings, and brooch, made by Viennese jeweler A. E. Köchert and often worn by Princess Isabella of Croÿ (1856–1931). The necklace was last seen in public on the décolletage of the late American comedian Joan Rivers at a Golden Globe Awards ceremony in 2004—apparently on loan from New York jeweler Fred Leighton.

The beautiful gem peridot is formed in the Earth's mantle and brought to the surface through volcanic action and movement in the planet's crust. It is found in Pakistan, Myanmar, Vietnam, Tanzania, the US, and China's Changbai Mountains, now the most important source of peridot. In the ancient world, it was mined on Topazius (today's Jazirat Zabarjad), a volcanic island in the Red Sea. Pliny the Elder recorded in his *Natural History* (77 CE) that peridot was the largest of all known

Peridot, diamond, and pearl brooch

> ❝ The pistachio-colored peridot. ❞

Oscar Wilde, *The Picture of Dorian Gray*, 1890

gemstones. In the Middle Ages, Christian crusaders returning from military expeditions in the Middle East took peridot back to Europe, where it was used to decorate Church vestments and objects.

Small peridot crystals have been found in stony-iron meteorites known as pallasites. Examples found in pallasites in Molong, Australia, in 1912, and in Esquel, Argentina, were just large enough to produce faceted gems. Pallasite gems are usually less than 1 carat, extremely rare, and do not appear on the gemstone market.

In the past, peridot was sometimes mistaken for emeralds. When used in jewelery, the stones were sometimes backed with foil to intensify their color. According to crystal lore, the gemstone is both protective and purifying, helping a person to unburden themselves of past negative feelings and be open to learning from experience.

Apatite

Madagascan first

The name apatite derives from *apatē*, the ancient Greek word for "deceit," apparently because its crystals come in so many different shades and forms that it is not easy to identify. Chemically, it is a calcium fluorine phosphate, widely distributed in different rock types around the globe. In pure form, it would be colorless, but traces of other minerals are invariably present, giving apatite crystals their spectrum of hues, from pale yellow to green, blue, and eye-catching purple. Many stones are fluorescent when exposed to ultraviolet light.

The most sought-after apatite variety is electric- (or neon-) blue Madagascar apatite, discovered, it is said, by a child in the sand near a former gold mine in Madagascar. With their bold color, these Madagascan stones were the first to become popular in jewelery, though many of the brightest stones on the market today have been heat-treated. Purple stones, rare and also prized, include the world's largest specimen—the Roebling Apatite, a 500-carat violet crystal from Auburn, Maine, the only place where purple apatites are commonly found, now on display in the Smithsonian Institution in Washington, D.C. The stone is named after its donor, engineer and amateur gem collector Washington A. Roebling, the builder of the Brooklyn Bridge.

> " The first spiritual want of barbarous man is decoration. "
>
> **Archibald Campbell Carlyle**

Apatite and gold
necklace, designed
by Lori Kaplan, 2022

Brazil produces sea-blue stones, which are said by crystal healers to help creativity. Yellow apatites, found extensively in Mexico, are believed to increase sensitivity toward the divine. Other apatite varieties include the so-called asparagus stones, named for their yellowish-green color, which were first discovered in southeastern Spain. Some green and blue-green stones from Sri Lanka and Myanmar display chatoyance—a cat's-eye effect, due to the way light is reflected and scattered within the stone.

Benitoite

◇

The heavenly stone

Rock containing benitoite and calcite

In the early 1900s, a gold prospector called James Marshall Couch came across some brilliant blue crystals near the San Benito River in California. The stones' intense color led him to think he had discovered sapphires, but they were later identified as a completely new mineral and named benitoite after the area where the crystals were found. The new gemstone quickly became popular for its high luster, and was marketed as "the heavenly stone" on account of its deep blue color. In 1985, benitoite was designated the gemstone of the state of California.

Although small deposits have been found in other places around the world, the San Benito mine was the only source of gem-quality benitoite. However, the deposit gradually became worked out, and the mine closed in 2005, making this gemstone extremely rare and expensive.

Benitoite's beautiful blue color comes from the presence of iron and titanium, but there are also pink and colorless versions of the gemstone. It is strongly dichroic (displaying different colors from different angles), and may appear an intense blue along the crystal's length and colorless from above. Its most appealing qualities, however, are its brilliance and fire—the flashes of light that make a gemstone sparkle. It works well faceted, but it is only moderately hard, so it is usually placed in protective settings. The largest-known cut benitoite is around 15 carats in size. Cut stones are usually small, and faceted stones above 1 carat are rare.

In the world of crystal lore, benitoite is known as a high-energy stone that increases joy and positivity. It is thought to be able to help in the release of negative past experiences in order to make way for new challenges. The stone is also associated with the third eye chakra (in the forehead), believed to expand consciousness and increase a person's intuition.

" Imagine the color of a light blue sapphire combined with the fire of a diamond. "

GemCollector.com

Iolite

The Viking compass

Blue iolites have sometimes been known as "water sapphires" on account of their violet blue intensity. In 1996, American gemologist W. Dan Hauser discovered a huge, 1,714-carat piece of gem-quality iolite at Palmer Creek, Wyoming. Christened the Palmer Creek Blue Star, it was the world's largest piece of iolite at the time. However, just a few years later the stone was surpassed by a much larger specimen, a 24,150-carat piece of iolite found at Grizzly Creek to the south of Palmer Creek.

Iolite is the gem-quality version of the mineral cordierite, and was officially recognized in 1813. Its name comes from the Greek word *ios*, meaning "violet," and its color ranges from blue-violet to blue. Its most unusual feature is its marked pleochrism—the ability to absorb different wavelengths of light depending on the angle of the crystals. A thin sliver of iolite can act as a filter to reduce the glare from the sun. According to legend, the Vikings used iolite as a navigation aid during their long sea voyages. The stone changes color in direct sunlight even on cloudy days, enabling their navigators to use iolite to track the sun's position.

The colors of iolite vary according to the stone's body color. Violet iolites display light and dark violet and yellow-brown hues. Blue iolites show light and dark blue and either pale yellow or no color. Those who believe in the supernatural power of crystals believe iolite to be a powerful vision stone, providing inner strength and clarity of thought, and dissolving discord with others. It is sometimes recommended as an aid to meditation, as it is thought to calm the emotions and still the mind. Iolite is also said to clear the third-eye chakra, thus releasing creativity.

Blue-violet iolite
in its rough state

" The glassy appearance of iolite is so peculiar that it can be confounded with nothing but blue quartz. "

James Dwight Dana, 1852

Kunzite

True love's token

In 1963, US President John F. Kennedy bought a ring for his wife Jackie to celebrate their tenth Christmas together. The ring consisted of a 47-carat kunzite—a delicate pink stone—surrounded by diamonds and set in gold. The story goes that Kennedy was assassinated before he had given her the ring, and that she would not be separated from it after his death. After her own death in 1994, the ring sold at auction for $415,000—50 times its estimated value.

Kunzite was first identified as a distinct gemstone in 1902, when some pink crystals were found in San Diego County, California, and sent to American mineralogist and gem collector George F. Kunz, who worked for Tiffany and Co. Kunz identified the crystals as a gem-quality, pink version of spodumene, a mineral mined primarily for its lithium content, which is used in batteries and medicine. The gem was named kunzite in honor of Kunz. The California mines are now nearly exhausted, but the gemstone is also found in Afghanistan, Pakistan, Myanmar, Brazil, and Madagascar.

Kunzite forms in large crystals. One of the largest known examples is an 880-carat crystal now in the Smithsonian Institute in Washington, D.C. The crystals range from pink to violet, the color determined by the level of manganese in the stone. The deeper and more intense the color, the more valuable the gem. Some stones display phosphorescence: they glow in the dark after exposure to the sun. However, the color can fade if a stone is left in the sun for too long. Some stones are also pleochroic—meaning that the color varies when the stone is turned.

Kunzite is said to be calming and mood-lifting, and to aid meditation. It is associated with the heart chakra, and therefore linked to love and compassion.

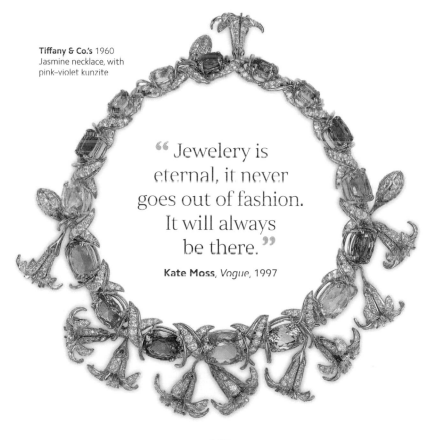

Tiffany & Co.'s 1960 Jasmine necklace, with pink–violet kunzite

> **"** Jewelery is eternal, it never goes out of fashion. It will always be there. **"**
>
> **Kate Moss**, *Vogue*, 1997

Malachite

Swirls of black and green

Polished malachite

First mined in around 4000 BCE in the Sinai area of the Middle East, malachite lies in deep underground caverns and fissures in the form of stalactites and rock coatings. It is often found over copper deposits, hence its abundance in the Copper Crescent of the Democratic Republic of the Congo. Removed in slabs, it is cut into slices to reveal the vivid, swirling, green-banded patterning for which it is renowned.

The ancient Egyptians used malachite for jewelery, and carved it into ornaments and amulets that were believed to ward off the evil eye. Ground to a powder, it is one of the oldest green pigments known to exist and was used for tomb paintings and eye makeup. In India, artists used malachite in mural paintings and illuminated manuscripts; in China, it was used to make inks. Painters in Renaissance Europe added it to oil paints.

Malachite is an opaque, relatively soft stone that is easy to carve and polish, making it ideal for cameos. It has also long been popular as a veneer for decorating furniture. In the 1800s, two immense malachite deposits were discovered in the Ural Mountains in Russia, and malachite soon became a favorite of the imperial family. The most spectacular display of malachite in the world is the Malachite Room of the Winter Palace in St. Petersburg. Designed in the 1830s for the wife of Czar Nicholas I, the room has malachite pillars, fireplace, and grand ornamental vases and urns.

In the past, malachite was worn to ward off evil influences. In crystal healing, it is considered to be a transformational stone that promotes emotional balance.

> ❝ Joséphine loved fantasy jewelery, and the malachite and pearl cameo parure was one of her favorites. ❞
>
> **Rapaport**, 2021

The Empress Joséphine's malachite cameo and pearl parure, c. 1810

Opals represent the tips of peacock feathers in a sapphire, peridot, and opal brooch, 1908.

Opal

Iridescent beauty

The opaque and changing nature of opals has inspired many tales and superstitions. It was said that a ninth-century Bishop of Rome called opals *patronus furum*—the patron stone of thieves. He believed that pickpockets wore opals to sharpen their own eyes while clouding the eyes of others. Opals were also regarded suspiciously during the Black Death (1346–1351). Some people claimed that the vibrant colors of opals were brought out by the sweat of high fevers but extinguished at the point of death when the body rapidly cooled.

> **"** Some by their splendor rival the colors of the painters, others the flame of burning sulfur or of fire quickened by oil. **"**

William Shakespeare, *Twelfth Night*, 1602

Opals have been a part of human culture for centuries. Carved opal beads from Kenyan burials date from 4000 BCE. In Europe, opals mined in Hungary were highly valued in the Roman Empire. Pliny the Elder wrote in his *Natural History* that only emeralds were more prized. He went on to say how counterfelt opals made of glass by unscrupulous craftsmen could be distinguished from true opals by their lack of flashing brilliance in sunlight.

Opals can be clear, translucent, or opaque; they can show all the colors of the rainbow or just one. There are two main types: precious opals, the most valuable, which exhibit flashes of color under a light source, and common opals, which lack iridescence. Formed from a mixture of silica and water in the cavities between rocks, opals are amorphous in structure—made up of millions of tiny spheres packed closely together without a pattern. This makes opals a delicate stone, susceptible to cracking through heat or the absorption of harsh detergents. The play of color in precious opals is produced by the diffraction of light within the stone.

Originally mined in Slovakia and Ethiopia, opals are now mainly mined in Mexico, Brazil, and Australia, with Australia accounting for 95 percent of the opals found today. The largest gem-quality opal ever found was the *Olympic Australis* at 11 in (28 cm) long. Opals feature in the origin stories

Different colored opals

of Australia's Indigenous people. The Pitjantjatjara people near Uluru, in the central Australian desert, associate opals with the Rainbow Serpent, a powerful creator said to have shaped the Earth with the undulating movements of its body. According to myth, when the Serpent finished creating the Earth, it moved between water holes, creating a rainbow in the sky. Any scales that fell from its body became opals.

The fascinating colors and changing lights of opals have earned them a place in some of the finest jewelery, even though their tendency to fracture make them difficult to carve. Opals were loved in 16th-century England, and Napoleon gave the Empress Joséphine a spectacular opal ring—now lost—called the Burning of Troy because of its fiery colors. Queen Victoria was also a devotee of the stone.

Opal jewelery was especially popular during the Art Nouveau period of the early 20th century. One of the most stunning opal pieces of this period is the "La Source" pendant created by René Lalique in 1902. Huge Mexican opals, known as fire opals for their vibrant orange color, form part of a bower for a classical ivory sculpture. Like much of Lalique's jewelery, "La Source," at 3 in (7 cm) long, is oversize by modern standards, owing, it is said, to Lalique regularly taking commissions from the French actress Sarah Bernhardt, who wanted her jewelery to be clearly visible to the audience when she was

Dinosaur opals

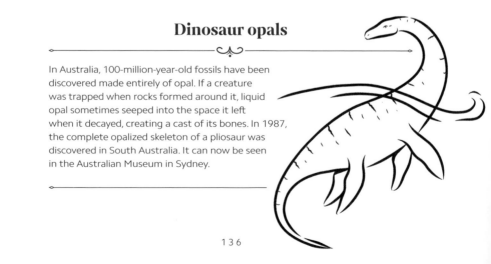

In Australia, 100-million-year-old fossils have been discovered made entirely of opal. If a creature was trapped when rocks formed around it, liquid opal sometimes seeped into the space it left when it decayed, creating a cast of its bones. In 1987, the complete opalized skeleton of a pliosaur was discovered in South Australia. It can now be seen in the Australian Museum in Sydney.

Cate Blanchett wears chandelier-style opal earrings by celebrity jeweler Chopard at the Oscars in 2014.

> " The hues of the opal, the light of the diamond, are not to be seen if the eye is too near. "

Ralph Waldo Emerson, "Friendship," 1841

on stage. Purchased by art collector Henry Walters at the 1904 World's Fair in St. Louis, Missouri, the pendant is now in The Walters Art Museum in Baltimore.

Opals lost some of their appeal in the late 20th century, when they gained a reputation for being unlucky if they were not your birthstone (they are the stone for October), but they have recently made a comeback. Among the celebrities who have been photographed wearing opals on the red carpet are Jennifer Hudson, Cate Blanchett, and Taylor Swift.

Spinel

The great impostor

Catherine the Great wearing the spinel-topped crown commissioned for her coronation in 1762.

A richly colored gemstone of great clarity and sparkle, spinel has often been confused with ruby. Large stones of great beauty were chosen to adorn the crowns of emperors and kings across Asia and Europe until the late 18th century. Many came from mines in the Badakhshan region in modern-day Tajikistan, which produced large specimens known as Balas rubies.

In 1783, French mineralogist Jean-Baptiste Louis Romé de l'Isle discovered that spinel has a different chemical composition from ruby, and a number of famous "rubies" had to be reclassified. The Black Prince's Ruby mounted on the front of the British Imperial State Crown, for example, is a 174-carat spinel, and the ruby that adorned the Imperial crown of Russia is a 412.25-carat spinel. Timur's Ruby, also in the British Crown Jewels, is a 350-carat

" Polished by the spirits. "

Burmese saying

The record-breaking Hope Spinel

spinel. Although the stone is named after the 14th-century Emperor Timur, founder of the Timurid Dynasty and conqueror of Delhi in 1398, Persian inscriptions listing six of its owners do not include Timur himself. The stone was presented to Queen Victoria following the colonization of India by Britain in the 1840s.

The name spinel probably derives from the Latin word *spina* meaning spine or thorn, due to the stone's spike-shaped crystals. Although it continued to be used for jewelery, spinel declined in value from the 1920s, as people preferred "real" rubies. Since the 1980s, however, its fortunes have revived. Its clear color and sparkle appeal to modern designers, and antique spinels are now highly sought after. In 2015, the Hope Spinel broke records when it fetched £962,500 at auction in London. Named after a former owner, 19th-century London banker Henry Philip Hope, it is a 50.13-carat plum-red spinel from the Kuh-i-Lal mines in Tajikistan, the source of many large spinels.

OTHER SEMIPRECIOUS GEMS

Spinel deposits occur in many parts of the world, though Myanmar, Sri Lanka, Thailand, Vietnam, Afghanistan, and recently Tanzania, are the main sources. Spinels are often found alongside rubies and other gemstones. Most gem-quality specimens are small, but larger pieces are also found. The Samarian Spinel, a large polished stone captured during Persia's 18th-century invasion of India, and now in the Iranian Crown Jewels, weighs in at more than 3½ oz (100 g). It is believed to be the largest red spinel ever discovered.

Spinels occur in a kaleidoscope of colors from pink, orange, and red, through plum and violet, to blue, and even black. Red is the most valuable, though any color should be clear and bright with good color saturation. Strong, bright blues, found in Vietnam and Sri Lanka, are highly valued.

In crystal healing, spinel is considered to be a high energy stone. Practitioners use it for the rejuvenation of body and mind and to connect with the natural world. It is believed to bring about emotional healing by reducing anxiety and stress. Because of the range of colors, it can revitalize every chakra in the body. Black spinel, for example, is said to unblock the root chakra, leading to a feeling of security. Blue spinel is believed to promote calm, orange creativity, and red to inspire passion and vitality.

The Côte de Bretagne

In 1749, the French king, Louis XV, joined the chivalric Order of the Golden Fleece. The court jeweler created a Rococo-style insignia for him using gems from the royal collection. At the center was the oldest gem, a Badakhshan spinel known as the Côte de Bretagne, carved into the form of a flame-breathing dragon. During the French Revolution (1789–1799), the insignia was stolen and broken up. Most of the jewels were never recovered, but the Côte de Bretagne is now on display in the Louvre Museum in Paris.

Tanzanite

Out of Africa

Prized for its brilliant blue color, tanzanite is a transparent
variety of the mineral zoisite—chemically, a calcium aluminum
silicate. Nowadays, it is an established favorite gemstone, and
mentioned in the same breath as diamonds, rubies, and emeralds.
But tanzanite is a late arrival in the jewelery pantheon. It was only
in 1967 that a Goan-born tailor and amateur prospector, Manuel de
Souza, came across the dazzling blue crystals in northern

> ❝ Tanzanite is the first transparent deep blue gemstone to be discovered in more than 2,000 years. ❞

Tiffany & Co., 1969

Tanzania's Merelani Hills, southwest of Mount Kilimanjaro, and spotted their commercial potential. New York jewelers Tiffany and Co. were early investors, naming the gems "tanzanite" after their country of origin.

Although tanzanite is considered to be a relatively affordable gemstone, its velvety blue color has caught the eye of celebrities and royalty, and its value and popularity have steadily increased. The largest tanzanite ever sold, a 423.56-carat specimen, modeled by American actor Sarah Jessica Parker before a charity auction in Hong Kong in 2016, sold for just over $350,000.

Tanzanite is one of the strongest pleochroic gems (able to absorb light waves of different lengths depending on the orientation of the crystals) and one of very few trichroic gemstones—displaying three colors within the same gemstone (blue, purplish-pink, and yellow). Most stones undergo heat treatment to remove the yellow, leaving the very desirable purplish-blue. The predominant hue will depend on the orientation of the gem when it was cut.

Another zoisite gemstone, also mined in Tanzania, is anyolite—sometimes called ruby-in-zoisite—in which chromium-rich green zoisite crystals exist in combination with opaque ruby crystals, providing a striking contrast. The Telemark region of southern Norway yields thulite (after Thule, an ancient name for Norway), an eye-catching, mostly opaque pink zoisite, first discovered in 1820.

A tanzanite in a triangular shape
known as trillion-cut

Tourmaline

The rainbow gem

This beautiful and popular gem occurs in a spectrum of colors—pink, red, purple, blue, green, brown, and black. Two or more colors can even appear in the same stone. Historically, tourmalines were often confused with other gemstones. In 1541, when Portuguese conquistador Francisco Spinoza came across bright green gemstones in the forests of Brazil, he thought he had found emeralds. Consequentially, large quantities were sent to the jewelery trade in Europe. Brazilian green tourmalines are still called "Brazilian emeralds" today.

Red tourmalines were often mistaken for rubies, and in the 1800s, Dutch traders imported a large collection of tourmalines from Sri Lanka into Europe in the belief that they were zircons. When the

Tourmaline
displaying "color zoning"

stones turned out to be a distinct and previously unidentified mineral, it was named tourmaline from the Sinhalese phrase *tura mali*, meaning "stones of many colors."

Tourmaline was discovered in the US in the 1820s, when two naturalists hiking on Mount Mica in Maine spotted green and red stones among the rocks. Nearly 50 years later, the first industrial gemstone mine in the US opened in the area, around the time that deposits of tourmaline were also discovered in Southern California.

American gemologist George F. Kunz championed Maine and Californian tourmalines, and one day in 1876, having "wrapped a tourmaline in a bit of gem paper" set off to persuade Charles Tiffany, founder of the New York jeweler Tiffany & Co, to include them in the company's designs. He

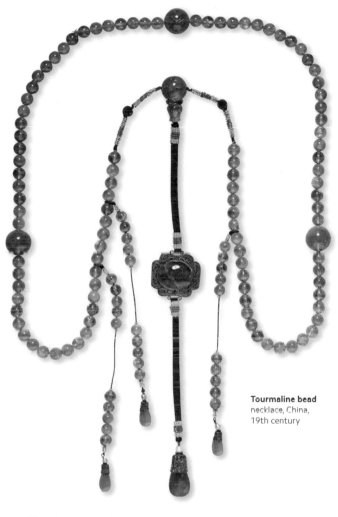

Tourmaline bead
necklace, China,
19th century

" The endless gorgeous colors of
these unexploited gems. "

George F. Kunz

Caesar's Ruby

The 225-carat, deep pink stone known as Caesar's Ruby is carved in the shape of a bunch of grapes, or a raspberry, and set as a pendant with green and gold enamel "leaves." It was given to Catherine the Great of Russia by King Gustav III of Sweden in 1777, having previously been in the collections of the French and then the Austrian royal families. In 1922, mineralogist Alexander Fersman was making an inventory of the Romanov jewelery collection when he discovered that the ruby was actually a rubellite tourmaline.

was successful. Kunz went on to source other colored gemstones for Tiffany as the company's chief gemologist. In the 1920s and '30s, tourmaline was popular in Art Deco cocktail rings.

The Chinese Empress Dowager Cixi (tenure 1861–1908) was another fan of the stone, especially red tourmaline, and imported large quantities from the Californian mines. It was set into jewelery or carved into snuffboxes. The empress's death put an end to the fashion for tourmaline in China, but trade in the gemstone expanded around the rest of the world. Extensive tourmaline deposits were found in Brazil, and smaller quantities in Afghanistan, Pakistan, Nigeria, and Madagascar.

Tourmaline is the name given to a group of related minerals that share the same crystal structure but differ in their chemical composition. Some contain additional elements, which give the stones their rainbow colors. There are more than 30 types, of which five—elbaite, schorl, dravite, liddicoatite, and uvite—mostly appear in the gem trade. Scientific analysis is required to tell the types apart, so for simplicity tourmalines are usually known by their color.

Elbaite, named after the Italian island of Elba, where it was found in 1913, is the most common type of tourmaline. It comes in a range

Rubellite tourmaline
pendant, China,
18th–19th centuries

Tourmaline brooch by René Lalique, c.1903–1904

of colors. In the past, the red variety, known as rubellite, whose deep pink or red hue is caused by traces of manganese, was often confused with rubies and spinels even though it is slightly pinker. Verdelite refers to elbaite that is yellow to bluish green, caused by traces of iron, while chromite is a bright, strong green that can resemble emeralds. The blue variety of elbaite is known as indicolite. Rarer than red or green stones, it occurs in a wide range of shades. Achroite is a colorless variety of elbaite.

The schorl and dravite types of tourmaline are dark brown or black, and opaque. Schorl can be completely black, whereas dravite always shows some brown. They are not often cut as gemstones, but schorl has been carved for use as mourning jewelery. Liddicoatite is from Madagascar and comes in pink, blue, and green. It is usually cut into slices and polished rather than faceted. Uvite is brown, dark green, or black, and comes from Sri Lanka.

> " Let me have these broken lights—these harmonies and dissonances of color. "

Oscar Wilde

In 1989, a new variety of tourmaline that was greenish-blue, caused by the presence of copper, was found in the state of Paraíba in northeastern Brazil. The exceptional clarity and saturation of Paraíba tourmalines soon caught the attention of jewelers, and the stone rapidly increased in value. In the early 2000s, similarly colored tourmalines were found in Nigeria and Mozambique. Although these are sometimes also referred to as Paraíba tourmalines, they are not as valuable as the Brazilian stones.

Some tourmalines display a range of colors in one crystal, a property known as color zoning. It occurs when conditions change as the crystal is forming, producing two or more colors either along the length of the crystal or layered on top of each other. It is an undesirable effect in other gemstones, but in tourmalines it is popular and valuable. Color-zoned stones are often cut into slices and polished. The watermelon variety, which has a red center and a green "skin" when sliced, is the most popular.

Another property of some tourmalines, especially indicolite, is pleochroism—the shade or even color of the stone changes when it is viewed from different angles. A darker color is seen when the stone is looked at along the length of the crystal.

Hidden properties

Heat can produce a positive and a negative charge in tourmaline. When the two charges connect within the crystal, they create a force that can attract tiny particles of dust, ash, or sawdust. The ancient Greeks were aware of this, and in the 18th century, the Dutch used tourmaline to extract warm ash from their pipes. They called tourmalines *aschentrekkers*, or ash drawers.

Pressure applied to tourmaline also generates an electric charge. For this reason, tourmaline has been used to detect variations in pressure and acceleration in a number of settings. In World War II, for instance, it was used in submarines to detect underwater explosions.

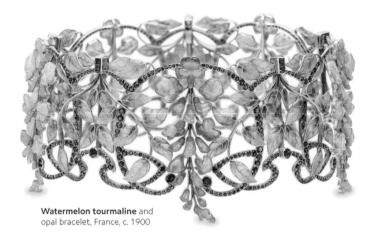

Watermelon tourmaline and
opal bracelet, France, c. 1900

> " A fine tourmaline answers
our idea of what a talismanic gem
or a gem-amulet should be "

George F. Kunz

In the world of crystal healing, tourmaline is considered to be a powerful stone with the ability to promote emotional and physical healing. It is said to improve confidence and understanding and encourage compassion and sympathy. Rubellite is thought to inspire love, creativity, and wisdom. It is also thought to be one of the most effective heart stones, promoting physical energy. Indicolite is known as the peace stone. It is said to encourage tolerance, truth, and harmony, and is the healer of past hurts and sadness. Verdelite (green tourmaline) is believed to stimulate the heart chakra, and inspire compassion and happiness. Schorl (black tourmaline) reputedly protects against negative forces. Watermelon tourmaline is said to aid emotional healing.

Turquoise

◈

The sky stone

Turquoise's history in Europe is apparent in its name. Coming from the French *pierre tourques*, meaning "Turkish stone," it refers to how turquoise was introduced by Turkish traders in the 11th century. But turquoise's history stretches back much further than this. It features in jewelery and ritual objects from ancient Mesopotamia and ancient Egypt (from around 3000 BCE), and some say that turquoise was the first gem humans ever mined.

Turquoise is usually a vibrant greenish blue, earning it the nickname "sky stone" among Indigenous peoples of America's Southwest. As an ancient talisman, turquoise has been said to possess many different spiritual qualities, but in modern crystal healing it is often associated with communication; it is also said to lift low spirits.

In ancient Egypt, turquoise was worn for protection, and often inlaid in gold amulets. The death mask of King Tutankhamen includes turquoise on the cobra and vulture adornments on the forehead. Turquoise was so important to the ancient Egyptians that their fertility goddess, Hathor, was also called "Mistress of the Turquoise" and was believed to watch over those who mined it.

Helmet of Suleiman the Magnificent, studded with turquoise and rubies, 16th century

“ Assuredly the turquoise
doth possess a soul more
intelligent than that of man. ”

Anselmus de Boodt, 1609

> **❝** Whoever wears a turquoise ring
> will never become dependent. **❞**

Hadith, Imam Ali al-Rida, eighth century

Turquoise was also held in high esteem in ancient Persia (559–331 BCE), now Iran. According to one Persian writer, Abu Rayhan al-Biruni, the stone was a symbol of "victory, strength, and protection from malefice." In fact, the Persian word for turquoise is *firuza*, while the word for victory is *firuzi*. According to a famous saying, good fortune could be obtained by seeing the new moon reflected on a friend's face, on a copy of the Quran, or on a turquoise.

For centuries, Persia dominated the trade in turquoise; the gems that reached Europe through Turkey largely originated here. Turkish riders would adorn their horses with the blue stone, believing that it would stop them from falling and keep the horses healthy. But turquoise also featured in the myths of cultures farther east. According to legend, it played a role in the union of Tibetan king Songtsen Gampo (r.618–650 CE) and Tang Chinese princess Wencheng (620–682). The Tang emperor did not want his daughter

Double-headed serpent

Nine turquoise mosaics now held in the British Museum are masterpieces of Mexica (Aztec) and Mixtec skill, undoubtedly taken to Europe by the Spanish conquistadors who colonized Central and South America in the late 15th century. Among the mosaics is a double-headed serpent, possibly worn as a chest decoration or affixed to a staff, made of hundreds of tiny turquoise pieces. To the

Mexica, the double-headed serpent, or *maquizcóatl*, could be an omen of death, but it also served as a reminder of the continuous cycle of life in Mexica religion.

A Tibetan *ga'u* (amulet case)
set with turquoise

to marry the Tibetan king, and proposed a challenge to decide who she would wed. He invited envoys from several kings to try to thread silk through a winding channel in a piece of turquoise the size of a shield. After other suitors failed, Tibetan Minister Gar tied the silk around an ant, put the insect near a hole, and gently blew on it until it walked through the channel to the other side, successfully threading the silk.

The craze for turquoise in Europe during the 19th century was led by Queen Victoria. She supposedly presented her friends cameo rings containing her own portrait, set in a ring of perfect turquoise cabochons (shaped and rounded, rather than cut) and gave majestic eagle-shaped turquoise, ruby, diamond, and pearl brooches to the 12 women who carried the impressive train of her wedding dress. Some of Victoria's favorite turquoise jewelery included a turquoise and diamond tiara and matching parure designed by her beloved husband, Albert, in 1843. After Albert's death in 1861, Victoria could not bear to wear such colorful pieces, and she gave this set to her daughter Princess Helena, as a wedding present in 1866.

Victoria also owned several turquoise brooches, including a pair given to her in 1838 by Louise of Orléans, Queen of the Belgians. These bow-shaped trinkets later became part of the jewelery collection of Valerie Eliot—the British wife of American poet T. S. Eliot—and in 2013 sold for over $10 million at auction. The Victorians tended to fashion turquoise into pavé patterns—many small pieces arranged closely together and set with other precious stones, such as diamonds. In the early 20th century, when turquoise became popular in Art Deco pieces, larger cabochons were more in vogue.

Turquoise has long been used by the Indigenous peoples of North America to make striking jewelery, often set in silver. For the Navajo people of the Southwest, turquoise is the "stone of life," a sacred rock said to bring good luck and good health. Because turquoise can change color, Navajo people look to these changes as indications of the wearer's health.

The Southwest US and Mexico are home to hundreds of turquoise mines, each producing its own kind of stone. One of the most popular

Navajo bird brooch

> ❝ It was my turquoise.
> I had it off Leah when I
> was a bachelor. ❞

William Shakespeare, *The Merchant of Venice*, 1605

The Marie-Louise Diadem

One of the most striking exhibits in the Smithsonian's National Museum of Natural History in Washington, D.C., is the Marie-Louise Diadem, a beautiful band of large turquoise stones and diamonds. But the diadem was not always like this. Its original owner, Empress Marie-Louise of France, the second wife of Napoleon, was painted several times in an emerald and diamond parure that included this diadem. After her death, the parure passed to a descendant, who sold it to Van Cleef & Arpels. Surprisingly, the French jeweler removed the emeralds, sold them, and inserted turquoise stones instead.

strains of turquoise comes from the Sleeping Beauty mine in Globe, Arizona, which gets its name from the unusually even pastel blue color of the stone. Sleeping Beauty turquoise has no "matrix"—the web of other colors that usually run through the gem. It was all the rage in the 1970s and '80s, when Italian buyers introduced it to Europe. Since the mine stopped producing turquoise in 2012, in order to focus on copper, the price for Sleeping Beauty turquoise has skyrocketed. It is now most commonly seen in high-end jewelery.

Turquoise continues to be a popular gem around the world, and many modern designers, such as Britain's Monica Vinader, still do as the Egyptians did, giving it a contrasting setting of bright yellow gold. Some of the most famous turquoise creations include Van Cleef & Arpels' 1974 Panka necklace containing 132 turquoise stones—worn by Eva Mendes to the Oscars in 2009—and Tiffany's turquoise and platinum bib necklace worn to the 2015 Oscars by Cate Blanchett. Van Cleef & Arpels uses Central and South American turquoise in some of its most exquisite work, from diamond-encrusted tiaras to whimsical lapel clips. But turquoise does not just feature in jewelry. In 2021, a Rolex Zenith Daytona sold for a staggering $3.1 million at auction, one of only two examples of the watch to be made with a turquoise face.

Zircon

Rock of ages

Zircon has been prized for millennia. The gemstone was said to form the leaves of the mythical Kalpavriksha, a divine tree of life composed of precious jewels, spoken of in Hinduism and other ancient Indian religions. It is the oldest mineral ever found on Earth. The fact that its crystals contain minute (non-dangerous) quantities of radioactive uranium, which decays at a known rate, allows scientists to pinpoint its age and understand the chemical processes taking place on Earth at the time.

The way zircon refracts and disperses light gives its crystals outstanding fire and brilliance. Pure zircon is made up of zirconium silicate, which is colorless, but trace amounts of other elements, such as zinc, calcium, iron, or titanium, produce many different colors: red, blue, green, and gold (yellow or orange). Its name derives either from *zarqun* an Arabic word for the color vermilion or *zargun*, a Persian word for "gold-colored." Traditionally, the term "jargoon" or "jargon"—also derived from *zargun*, has been applied to light yellow and colorless zircon stones, while the terms "hyacinth" or "jacinth" are applied to the darker, redder stones.

Zircon can never rival diamond, which is much harder and less brittle, but its affordability makes it an attractive gemstone for modern jewelery. Shimmering sea-blue zircons are the most popular, but the color in almost all of these has been artificially produced. Although blue zircons are sometimes found in nature, they are very rare: most blue stones on the market start

Faceted zircon

" No other species can show
such magnificent stones of
a golden-yellow hue. "

G. F. Herbert Smith, *Gemstones*, 1940

out as reddish brown crystals and are then heat-treated to turn them blue. Similarly, colorless stones—misleadingly called Matura or Matara "diamonds," after the once-important gem trade port of Matara in Sri Lanka—are also heat-treated.

As a mineral, zircon occurs widely around the globe, but the bulk of high-quality crystals, good enough to be cut as gemstones, come from Sri Lanka, Myanmar, Cambodia, and Vietnam; they are also found in Australia, New Zealand, southern Norway, and Quebec.

Organic Gems

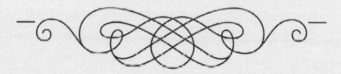

Amber

Gold of the north

Amber amulet, 1st century CE, The Netherlands

This much-prized gem, loved for its warm, rich color, is not really a gemstone at all but the fossilized resin of coniferous trees that lived millions of years ago. It has long been used for jewelery and decorative items, and as a talisman.

Amber is found all over the world, but the area around the Baltic Sea in northern Europe has always provided the most sought-after gems. Vast forests grew here in prehistoric times, creating the largest amber deposits on the planet. Other sources of gem-quality amber are the Dominican Republic and Myanmar.

The fossilization process took many millions of years. When the trees exuded resin, the sticky substance trapped insects, animals, and plants as it hardened. In all, nearly 7,000 species have been identified within amber, some from 130 million years ago, providing scientists with snapshots of prehistoric life.

Pliny the Elder's *Natural History* (77 CE) recounts some of his predecessors' ideas on the origins of amber. The Athenian politician Nicias (c. 470–413 BCE) suggested that the sun's rays left "an unctuous sweat on the earth's surface" that was swept into the sea and turned into amber. The Greek geographer Pytheas (c. 350–306 BCE) thought that it was "an excretion of the sea in concrete form." Pliny himself was aware of amber's real origins as a substance exuded by pine trees, and that it had been found in Egypt and India as well as in northern Europe.

Amber appeared in written records in China around 200 BCE. In Chinese mythology, it is called *hu po*, "the soul of the tiger," as it was believed the souls of dead tigers entered the earth and turned into amber. It was thought to have protective and healing powers due to its warm color and ability to retain heat.

Amber can be transparent or opaque. The colors range from yellow to deep golden-orange and, less often, red and brown. Its color darkens over time, and some people prefer the "antique" look that this gives it. Blue and green versions also occur, although a very bright yellow-green is likely to be imitation amber.

Because amber is lightweight and warm to the touch, it is comfortable to wear and suitable for large pieces of jewelery. A soft stone, it is not normally faceted. Instead, it is carved into beads

" With amber bracelets, beads, and all this knavery "

William Shakespeare, *Taming of the Shrew*, 1592

3rd-century CE amber necklace, Germany

> " Eurymachus received a golden necklace ... set with beads, that glowed as if with sunshine. "

Homer, *The Odyssey*, c. 8th–9th century BCE

and cabochons, or just polished. In 18th-century Italy, amber was carved into detailed reliefs and statues, usually on a religious theme. At Christmas, elaborate nativity scenes (*presepe*) were fashioned from amber.

Natural inclusions increase the value of amber. Insects, spiders, small frogs and lizards, bacteria, even a dinosaur feather, and plant and flower fragments have all been found within the gemstone. These living creatures were alive and struggling when they were encased in resin so they often appear in awkward positions, with little bubbles and cracks around them. If an insect appears neatly laid out and a little too perfect, it can be a sign that the amber may be imitation. Heat treatment can dispel cloudiness in amber, but it may also cause little circular marks known as "sun spangles" to appear. The effect is sometimes created deliberately.

In gem lore, amber is considered a powerful stone due to its great age, and has always been associated with wisdom. It is also regarded as a purifying stone, removing emotional and physical pain, and releasing positive energy and confidence. On a physical level, it is thought to cure headaches, colds, and flu, and to help the digestive system.

Amber pendant containing a spider and cricket, 19th century

The Amber Room

In the mid 1700s, a spectacular Amber Room was created for Catherine the Great, Empress of Russia (r. 1762–1796) in the Catherine Palace near St. Petersburg. Six tons of Baltic amber were used to decorate the room, either applied over gold leaf to wooden wall panels or carved into embellishments. Enchanted by the golden glow given off by the amber when lit by candlelight, visitors described the room as the Eighth Wonder of the World. When the German army entered St. Petersburg in 1941, soldiers dismantled the room and shipped the amber to a hiding place in Königsberg (modern Kaliningrad). It was never seen again.

Jet

◆————◆————◆

Death becomes it

Princess Louise, daughter of Queen Victoria, wearing mourning clothes, including jet jewelery in 1871

When Britain's Queen Victoria (r. 1837–1901) wore jet mourning jewelery following the death of her husband Prince Albert in 1861, she started a fashion that lasted for decades.

A Victorian jet and pearl mourning brooch containing the hair of the deceased

This lustrous black stone—from which the term "jet black" comes—is not a precious gemstone mined from the depths of the earth, but the fossilized remains of coniferous trees that lived during the Jurassic period, around 180 million years ago. Treasured for its rich black color and opacity, jet is a type of carbon-rich lignite related to coal. When trees died and fell into water, their remains became waterlogged and sank into the mud, where they were gradually covered in sediment. Over millions of years the layers of sediment built up and were subjected to intense pressure and heat, producing nodules and small seams of fossilized material

> " [Whitby jet is]
> Black and like a jewel. "
>
> **Gaius Julius Solinus,** c. 3rd century CE

Jet has been used for jewelery since the prehistoric period. Early examples of beads and carvings dated to around 10,000 BCE have been found in France and Germany, and beads, carvings, buttons, and amulets have been discovered in Bronze Age burial mounds (2500–2000 BCE) all around the UK. In 2019, archaeologists uncovered a set of 122 carved jet beads in a 4,000-year-old burial mound on the Isle of Man in the Irish Sea.

Jet is found in many countries, including Germany, Poland, Turkey, and the US. The most significant deposits occur in Spain, along the country's northern coast, and in the UK. The area around Whitby on England's Yorkshire coast has long been renowned for providing the best-quality jet in the world. Whitby jet formed from forests containing a high proportion of a now-extinct species of *araucaria*, or monkey puzzle tree, and was subjected to particularly high temperatures during the fossilization process. The deposit, which consists of nodules and narrow seams, runs along the coast and under the sea, and people used to find large quantities washed up on the area's beaches. The jet is particularly smooth in texture, making it easy to carve. The largest piece ever found is an entire monkey puzzle tree trunk calculated to be 180 million years old and measuring 21 ft (6.4 m) in length.

Jet was popular with the ancient Romans, who used it for jewelery and its supposed medicinal properties. They called it *gagates* after the ancient town of Gagae in Lycia, southwest Turkey, from where they obtained the bulk of their supplies. In Roman Britain, it was collected from the beaches around Whitby. At the time, the nearby city of York was full of jet workshops turning out amulets, necklaces, hairpins, and other items to be traded throughout Europe. Pendants carved with the head of Medusa, one of the three Gorgons of Greek mythology, were particularly popular as talismans. The Romans also believed that smoke from burning jet had medicinal benefits.

Jet reached the height of its popularity in 19th-century Britain, and featured in the Great Exhibition of 1851. Following Albert's death, Queen Victoria's mourning jewelery made from Whitby jet sparked a fashion around the world. Other items that became popular included carved jet letter-openers, rosaries, combs, and seals. Demand for jet was so high in Britain during the 1860s and '70s that supplies on the beaches

Victorian jet brooch

Clenched fist

Since prehistoric times, people around the world, from ancient Greece and Rome to India, have worn jet amulets as a protection against the evil eye. In 16th century Spain, Spanish jet, or *azabache*, carvings were popular religious talismans, and were associated with the Camino de Santiago pilgrimage route across the north of the country. A jet clenched fist, or *mana figa*, in particular, was believed to protect a pilgrim against disease and violence on their journey.

around Whitby ran low and a mining industry sprang up in the area to meet the demand. People tunneled into the cliffs to reach the thin seams that ran along the coast—a practice that continued until the 1920s, when it was banned on safety grounds—and additional supplies of jet were imported from northern Spain to be turned into jewelery in Whitby. Jet jewelery continued to be produced in England, but its appeal began to dwindle after Victoria's death in 1901, when colorful gems became popular again.

Most jet is smooth textured with a velvety luster, but some examples have a grainy texture resembling wood. Its even texture makes it suitable for precise, detailed carving. One test of genuine jet is the sharpness of the carving, as most substitute materials

> " I wore a black satin dress with embroideries & jet, a lace veil of old point, & diamond diadem & ornaments. "

Queen Victoria, 1896

A jet choker by Christian Dior, worn on the catwalk in 1999

" Favours from a maund she drew
Of amber, crystal, and beaded jet. "

William Shakespeare, *A Lover's Complaint*, 1609

cannot be carved with the same degree of precision. Jet can look attractive with a matte finish, but it also comes to a high polish— so much so that in past centuries, people used small pieces of jet as mirrors. It is lightweight, warm to the touch, and comfortable to wear. It can be cut en cabochon or as beads, or faceted (rose cuts are popular), and can be carved into brooches, cameos, and ornaments. Because of its relative softness, jet should be stored away from other jewelery so it does not get scratched, and should be wiped with oil occasionally so that it does not dry out and crack. When rubbed, jet produces a static electric charge that attracts tiny particles such as dust.

In gem healing, jet is a purifying and protective stone. It is related to the root chakra (at the base of the spine); its connection to the earth is purportedly good for grounding and balancing the mind and body. Some people also believe that it draws out negative energies and converts them into positivity, breaking damaging patterns of behavior.

Is it genuine?

Several materials have been used to produce imitation jet. During the height of its popularity in the 19th century, the naturally occurring black volcanic glass obsidian was often substituted, as was a manufactured black glass known as "Paris Jet," which tends to contain tell-tale bubbles. Scottish canel (or candle) coal—a wax extracted from coal seams—has also been tried.

A number of tests exist to help distinguish genuine jet. Being an organic gem, jet is warm to the touch in comparison with its imitators. And when a piece of genuine jet is rubbed against unglazed porcelain, it leaves a brown or black streak, known as a "streak test." A needlepoint test can also be performed—when the point of a heated needle is applied to genuine jet, it will give off an oily smell; if it is plastic, or rubber, however, it will smell acrid.

Coral

Red gold

Italian Woman in Coral Necklace, by Johann Heinrich Hasselhorst, 1863

This highly valued gem, sometimes referred to as "red gold," is the product of a tiny marine creature, the coral polyp. According to the Kumulipo, or Hawaiian Creation chant, the coral polyp was the first creature to appear on Earth. The polyp produces carbonate secretions that harden into delicately branching, tree- or fan-shaped structures, providing shelter for both the polyps and a host of other marine animals. The desecration of coral reef due to warmer oceans, pollution, and over-harvesting has led to coral trade restrictions in many

Branches of red coral
before it is cut
and polished

parts of the world. Following the example of US jeweler Tiffany & Co. In 2004, some jewelers have stopped using coral in their products.

There are many different colors, but the slow-growing red coral (*Corallium rubrum*) is considered the most precious, and the darker the shade, the more valuable it is. Coral polishes well, but is quite soft, so it is usually cut into cabochons, beads, and cameos. It has been prized for its beauty since ancient times. Coral amulets to protect the dead have been found in Neolithic graves (dating from c. 8000 BCE) in Switzerland, and coral was crafted into jewelery in ancient Egypt and Sumeria, from c. 3500 BCE.

Red coral is found in the Mediterranean Sea, Red Sea, Malay Archipelago, Pacific Ocean, and the seas around Japan. It was probably first harvested in the Mediterranean area and traded eastward by sea to India and southeast Asia, and southward across the Sahara. In many ancient cultures, it was associated with the planet Mars and

CORAL

" Born was the coral polyp,
born was the coral. **"**

Hawaiian creation myth

171

regarded as a symbol of protection. In the Hindu scriptures, Mars is called Mangal, and coral is known as Manga or Moonga. In China, coral represented longevity, and in Japan it was believed to bring good fortune and wealth. The Ancient Greeks linked coral's creation to the myth of Perseus and Medusa, the snake-haired Gorgon, and named it *Gorgonia nobilis*. According to the myth, after Perseus had slain the Gorgon, he laid her severed head on the seabed, where her blood turned the seaweed red and her fading gaze turned it into coral. The Greeks thought coral had the power to counteract witchcraft and poisons, and to protect ships against storms. In ancient Rome, children wore coral necklaces to ward off illness.

> " The Gorgon's poison in their spongy pith, they hardened at the touch. "

Ovid, *Metamorphoses*, 8 CE

West African societies acquired coral through their trade with European and Arab traders, and associated coral jewelery with high social standing. People of the Navajo, Hopi, and Pueblo nations of North America used it in jewelery after its introduction by European settlers.

In the 16th century, the Italian town of Torre del Greco, south of Naples, became a center for the coral industry. The first jewelery and carving factory opened in 1805, and many more followed. There is little coral to harvest in the area today, but the jewelery industry continues, using coral from Japan. Even here, however, coral is under threat. In 2017, the Japanese government declared reefs around the Japanese Ogasawara Islands, once a good source of red coral, extinct. Restrictions have been introduced for coral harvesting in Mediterranean waters, which specify a minimum size and a minimum depth for the coral being collected.

Silver and coral necklace, Algeria, 19th century

Protective measures

Eight species of red coral are identified in the coral trade. Four species need certificates of origin in order to be traded internationally, to show that they have been harvested legally. Until recently, customs authorities had to rely mainly on color alone to tell the species apart. Scientists have recently developed a genetic test, known as Coral-ID, that can distinguish species more accurately.

Ivory

White gold

The decimation of the world's elephant population has led to a global ban on all but a tiny section of the ivory trade. However, for centuries, the human love of ivory and the way it could be carved into objects of great beauty and delicacy, drove a highly profitable industry. Archaeologists have found carved ivory in the form of dice, combs, and little figurines of people and animals during excavations of Mohenjo-daro (now in Pakistan), one of the world's first cities, built around 2500 BCE.

> **"** Full gently now
> she takes him by the hand,
> A lily prison'd in a gaol of snow
> Or ivory in an alabaster band ... **"**

William Shakespeare, *Venus and Adonis*, 1683

Ivory is the dentin of tusks, which are effectively large teeth. Many mammals have tusks, including hippos, walruses, and killer whales, but the ivory trade has mainly centered on the tusks of elephants. African and Indian elephants produce different ivories. All African elephants have tusks, while only some male Indian elephants do, and African tusks are generally longer, less dense, and whiter in color. Indian elephant tusks, being more brittle, are harder to carve and yellow with age. It is thought that the ivory found at Mohenjo-daro came from Indian elephant tusks traded to central Asia, but ancient India

Ivory elephant decorated with sapphires, rubies, emeralds, and gold, c 1900

Mughal elephants

Elephants were especially revered in Mughal India, not only for their tusks but as a symbol of awe-inspiring strength and status. They were described by one 16th-century Indian poet as "waves of the ocean of calamity." Carved ivory elephants became a popular ornament in India, with artisans attempting to outdo each other in creating intricate, bejeweled creations. These trinkets would be decorated just as the Mughal elephants were in life, with gem-encrusted headdresses, carriage seats, and saddle blankets, and tinkling bells at their necks and feet.

" Only elephants should wear ivory. "

African Wildlife Federation

had begun to buy African ivory by about the sixth century BCE. Other ancient civilizations also made objects from ivory, including Egypt (3100 BCE–664 BCE) and Shang China (c. 1600–1050 BCE). Thutmose III's royal scribe recorded that the pharaoh hunted 120 elephants for their tusks on the way back from a military campaign in Syria.

Trade in ivory accelerated in the Roman Empire, when it signified wealth and status. Great quantities were shipped from North Africa. The Roman writer Pliny the Elder noted that "an ample supply of tusks is now rarely obtained except from India, all the rest of the world having succumbed to luxury." Ivory amulets were worn for protection or to increase fertility. In 1901, the excavation of a grave from the 4th century CE in York, England, revealed the remains of a woman originally from North Africa. Among her possessions were ivory and jet bangles, testifying to both her high status and the cosmopolitan nature of the Roman empire.

Ivory's link with fertility was also made in India. For both Sikh and Punjabi Hindu brides, an ivory bangle formed part of the *chooda*—a set of bangles worn during the wedding and for a period afterward.

In some societies, ivory signified holiness and chastity. Some of the most exquisite ivory objects are reliquaries and small figures of the crucified Christ made in Europe during the 12th century. At this time, ivory was in short supply and so reserved for the most precious work. As supplies increased in the 13th and 14th centuries, larger items were

Carved ivory box, made in Córdoba, Spain, during the reign of Abd al-Rahman III (c. 912–961 CE)

ORGANIC GEMS

Benin ivory masks

Oba Esigie, king of the 16th-century kingdom of Benin (in present-day Nigeria) carried an ivory pendant mask on his hip during ceremonies. The mask was carved to look like Idia, the mother of Oba Esigie, who created the title "queen mother" to honor her role in his life.

A symbol of Olokun, god of the sea, ivory was revered in Benin. It brought the kingdom great wealth, as it attracted Portuguese traders. Four related Idia masks, looted by the British when they invaded Benin in 1897, have survived to this day, all of them in collections outside Nigeria. Requests for their return have yet to be met.

made, including intricately carved religious scenes. In the 19th century, European colonization of Africa was partly driven by the demand for ivory, which was dubbed "white gold." The Republic of Côte d'Ivoire—Ivory Coast—in West Africa was named by French and Portuguese traders and colonizers for the large amounts of ivory traded through its ports, though East and South Africa were also active in the trade. This ivory was turned into carvings, knife handles, billiard balls, and, above all, piano keys—pianos being the latest must-have for Europe's expanding middle class. By the mid-20th century, the number of elephants in Africa had been reduced from 26 million in 1800 to fewer than one million.

In 1989, the Convention on International Trade in Endangered Species of Wild Fauna and Flora banned the ivory trade around the world, securing more than 100 nations as signatories. In many countries, the ivory ban applies not only to raw ivory but to ivory objects, even antiques, and a permit is needed to import and export them. From the 1970s, piano makers began to phase out ivory keys, and pianos are now made using synthetic alternatives. For most modern pianists, "tickling the ivories" doesn't actually involve touching real ivory.

Pearl

Pure luster

Portrait of Russian Empress Alexandra Feodorovna (1798–1851)
wearing pearls

❝ I must go seek some dewdrops here
And hang a pearl in every cowslip's ear. ❞

William Shakespeare, *A Midsummer Night's Dream*, c. 1596

ORGANIC GEMS

Known variously around the world as the queen of gems and the tears of the gods, pearls are a symbol of wealth and luxury. Many ancient cultures thought they were pieces of the moon, a belief reinforced by the moon's association with water, where pearls are found. In India, ancient texts describe the pearl as the moon's daughter. In Greek mythology, the pearl is linked to Aphrodite, the goddess of love and beauty, who rose from the sea.

Roman author Pliny the Elder states in his *Natural History* that "the first place and the topmost rank among all things of price is held by pearls." He recounts how Cleopatra owned two of the largest pearls known at the time. Wanting to impress her Roman lover Mark Antony with the extent of Egypt's wealth, she made a bet that she could provide the most lavish banquet. When the Roman general arrived, she dropped one of her pearls in a glass of vinegar and drank it. Pliny valued Cleopatra's two pearls at 60 million *sestertii*—$28.5 million today.

In the 16th, 17th, and 18th centuries, European rulers were often depicted wearing long strings of pearls as symbols of political power and wealth. Queen Elizabeth I of England (r. 1558–1603) was described as wearing "bushels of pearls," including thousands stitched onto her dresses. In Mughal India, where rulers effectively wore their treasury upon their bodies, 17th-century French traveler François Bernier described the relatively austere Emperor

Dragons and pearls

Chinese art often depicts dragons carrying pearls in their mouths, claws, or under their chins, or reaching out to seize a pearl from the air. The images refer to a myth about a greedy emperor whose people are afflicted by famine. A boy digging in his family's fields finds a pearl, which he hides in an empty rice sack. Although the fields die, the sack magically overflows with rice, news of which soon reaches the emperor. He summons the boy, who swallows the pearl and turns into a fire-breathing dragon, which protects the land and its people.

Aurangzeb (1618–1707) wearing "a necklace of immense pearls suspended from his neck that reached his stomach." At the same time, pearls became associated with love, fidelity, purity, and mourning. In 19th-century Europe, cameos and pearl chokers became popular, as did *ferronières*, headbands with a jewel suspended over the forehead, a fashion originating in medieval times.

Pearls are found in tropical and semitropical waters, particularly in the Arabian Gulf, the Gulf of Mannar between Sri Lanka and the southern tip of India, the Pacific Ocean, the coastal waters of East Africa, and the Red Sea. They vary in size from a millimeter or two across to the size of a pigeon's egg. One of the largest, the irregularly shaped Hope Pearl, now in London's Natural History Museum, is 2 in (5 cm) long.

Most natural pearls form in saltwater oysters belonging to the genus *Pinctada*. Traditionally, they were harvested from the seabed by free divers, who would dive more than 100 ft (30 m) for seven minutes or even longer using just one breath. This perilous occupation has mainly died out, but

The Hope Pearl, one of the largest known pearls

The oldest pearl

For a long time, the 5,000-year-old Jomon pearl found in Japan was thought to be the world's oldest known pearl. However, in recent years, pearls have been uncovered in a number of Neolithic burial sites dating from 5500–4000 BCE around the Arabian Gulf. The oldest pearl so far discovered is a 7,500-year-old pearl just 0.07 in (2 mm) in diameter found in sand adhering to a skull at a burial site at Umm al Quwain, in the UAE. The purpose of pearls in ancient burials is unknown, but one theory is that they were attached to the lips of the deceased as a burial rite.

" Pearls don't lie on the seashore.
If you want one, you must dive for it. "

Chinese proverb

Cultured pearls

Today, cultured pearls are produced by inserting a tiny bead into a mollusk, usually an oyster or mussel. The mollusk is then returned to the oyster beds and left for six months to two years to allow the pearl to grow. The size and color of cultured pearls are determined by the species of mollusk. Akoya pearls, from the *Pinctada fucata* oyster, are small and usually white, but can  be cream or pink. Tahitian black pearls, which grow in the large, black-lipped oyster, *Pinctada margaritifera*, are larger and come in dark browns, grays, and purples. The *Pinctada maxima* produce the largest pearls, in white, silver, or cream, occasionally reaching sizes of ¾ in (20 mm) in diameter.

it continues in some parts of the world, including among Japan's *ama* ("women of the ocean")—female free divers who practice a 2,000-year-old tradition.

Ancient myths tell of pearls being created when dragons fought in the clouds, or when raindrops fell into open oysters. In reality, a pearl forms when a piece of grit or a microscopic creature gets inside a mollusk and lodges in the soft tissue. To deal with the irritation, the mollusk secretes nacre, or mother-of-pearl—the same substance that lines the shell—around the intruder, sealing it inside a sac. The mollusk continues adding layers of nacre, and the pearl slowly grows.

Pearls can also form in mussels, clams, and conchs. The largest pearl ever discovered, known as the Pearl of Allah or the Pearl of Lao Tzu, was discovered in a giant clam in 1934. Freshwater pearls are found in rivers; well-known sites include the Mississippi in the US and the Spey and Tay rivers in Scotland. These pearls grow in mussels, often forming an elongated shape known as baroque.

The Chinese were the first to produce cultured pearls. For centuries, they placed tiny beads into oysters to deliberately produce a pearl. In the 1890s, Japanese noodle-seller Mikimoto Kōkichi started producing cultured pearls on a commercial basis. "My

La Pelegrina, suspended from a pearl, ruby, and diamond necklace

dream," he said, "is to adorn the neck of all women around the world with pearls." X-rays show the difference in internal structure between natural and cultured pearls. Natural pearls are made up of concentric rings, like an onion, whereas cultured pearls have parallel layers across the inserted material. Iridescence is produced when the overlapping layers diffract light.

Natural pearls are far more valuable than cultured ones due to their rarity. Other features that affect a pearl's value are its shape and texture—the rounder and smoother, the better—and its light-reflecting ability, or luster. This attribute has entranced humans for millennia. In ancient India, pearls were believed to have the power to prolong youth, cure eye diseases, and were prescribed as an antidote for poisoning. People in the Middle Ages thought pearls could restore strength, calm heart palpitations, and stop hemorrhages. In modern crystal healing, pearls are believed to relieve digestive disorders, allergies, and bronchitis, and to bring calm.

One of the most famous jewels in the world—a pearl known as La Pelegrina, or The Wanderer—was owned by Elizabeth Taylor. It is a 56-carat pearl given to her by

" I had just received La Pelegrina from New York ... I was touching it like a talisman. "

Elizabeth Taylor, 1969

Richard Burton. In the 1500s, it had belonged to the wife of Philip II of Spain, and then passed into the possession of the queens of Austria and France. At one time, it was owned by Napoleon Bonaparte. Taylor had it set in a necklace with rubies and other gems. The piece sold for over $11 million at the 2011 auction of her jewelery collection. Other famous pieces of pearl jewelery include the Baroda necklace, seven strands of flawless pearls commissioned by the Maharaja Khanderao Geakwad of Baroda, a state in what is now Gujarat, between 1856 and 1870. At some point reduced to two strands containing 68 pearls of graduating size, the necklace sold at auction in 2007 for more than $7 million.

In the 20th century in Western cultures, pearls represented a classic style—the equivalent of the little black dress. Fashion icons such as designer Coco Chanel, American First Lady Jackie Kennedy, and actors Audrey Hepburn and Grace Kelly were all regularly seen in pearls. In response to the claim that a woman did not feel dressed without her pearls, American writer Dorothy Parker quipped: "When I'm cold I just put another rope of pearls on."

The great pearl robbery

In 1913, London gem trader Max Mayer left a necklace of 61 pearls valued at £150,000 (around £17 million/$21 million today) with a potential buyer in Paris. Not wanting to purchase it, the client returned the necklace to Mayer by registered post. But when the gem trader received the package, all he found inside were sugar lumps. Following an investigation, the

British police arrested a well-known gang and recovered the necklace, but three pearls were missing. A little later, a member of the public was walking home when he saw a man deliberately drop a small package into the gutter. Curious, he opened it to find the three missing pearls. He handed them in to the police and received a reward of £10,000 from Lloyds Underwriters, which had insured the pearls.

Fashion designer Coco Chanel was often photographed wearing pearls, as here in 1938.

" A woman needs ropes
and ropes of pearls. "

Coco Chanel

Birthstones

❦

Each of the major gemstones is associated with a month of the year, and jewelery containing the stone is often given to people born in that month. In medieval Europe, when gems were associated with astrological signs rather than months, and widely thought to have mystical powers, wealthy people sometimes owned a complete set of 12 gemstones, which they wore in turn.

January: Garnet
The traditional birthstone for January, garnet is linked to spiritual strength and physical courage. Red garnet is associated with passion.

April: Diamond
The hardest and most brilliant gemstones, diamonds symbolize purity and commitment. They are said to bring inner strength.

February: Amethyst
Linked to royalty and wine, amethyst was historically believed to provide protection and promote healing. Warriors sometimes carried it into battle.

May: Emerald
Befitting its deep green color, emerald symbolizes rebirth. Traditionally, it was said to refresh the eyes and promote youth and vitality.

March: Bloodstone, Aquamarine
Originally represented by bloodstone, which is associated with the sun, March is also linked to sparkling aquamarine, a stone said to bring calm.

June: Pearl, Alexandrite
Although June is traditionally linked to pearl, representing beauty, modern lists include alexandrite, which is linked with good luck, as an alternative birthstone.

July: Ruby
"Blood drops from the heart" describes the ruby, the birthstone for July. It is associated with love, health, and wisdom, as well as strength in battle.

August: Peridot
Often used in chalices and crosses, peridot is said to protect the wearer from harm and to bring love, truth, and happiness.

September: Sapphire
The color of the sky, blue sapphire has long been associated with the divine, and is said to protect against the evil eye.

October: Opal, Pink Tourmaline
Known as the "magician's stone," opal is sometimes thought to be unlucky. This led to pink tourmaline, said to bring confidence, becoming an alternative stone for October.

November: Topaz, Citrine
Topaz is the original birthstone for November, but citrine was added to lists in the 1950s. Both are said to bring calm.

December: Turquoise, Blue Zircon, Tanzanite
While turquoise is the traditional stone for December, modern lists also include blue zircon and tanzanite. All three stones are said to provide protection when worn.

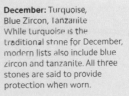

Index

Acknowledgments

Toucan Books

Editorial Director Ellen Dupont; **Editor** Dorothy Stannard; **Designer** Dave Jones; **Picture Researcher** Sharon Southren; **Authenticity Reader** Kit Heyam; **Proofreader** Julie Brooke; **Indexer** Marie Lorimer

Consultant Helen Molesworth
Writers Helen Douglas-Cooper; Lily Faber; Autumn Green; Andrew Kerr-Jarrett; Abigail Mitchell

Picture Credits

The publisher would like to thank the following for their kind permission to reproduce their photographs:

(Key: a-above; b-below/bottom; c-center; f-far; l-left; r-right; t-top)

11 Fotolia: apttone. **12 Alamy Stock Photo:** World History Archive. **13 Dorling Kindersley:** Ruth Jenkinson / Holts Gems (tc, tr, c, br, bl). **14 Alamy Stock Photo:** ScreenProd / Photononstop. **15 Getty Images:** Bettmann / Contributor. **16 Getty Images:** Clive Brunskill / Staff. **18 Bridgeman Images:** Photo © Christie's Images. **19 Shutterstock.com:** Granger. **20 The Metropolitan Museum of Art:** Purchase, Lila Acheson Wallace Gift, Acquisitions Fund and Mary Trumbull Adams Fund, 2015. **21 Bridgeman Images:** Photo © Christie's Images. **22 Bridgeman Images:** Photo © Christie's Images. **24 The Metropolitan Museum of Art:** Purchase, Harris Brisbane Dick Fund and The Vincent Astor Foundation Gift, 1984. **25 Getty Images:** National Galleries of Scotland / Contributor / Hulton Fine Art Collection. **27 Getty Images:** DEA / A. DAGLI ORTI / Contributor. **29 Shutterstock.com:** Kin Cheung / AP. **31 © Cartier:** Marian Gérard, Cartier Collection. **32 Alamy Stock Photo:** Artefact. **33 Bridgeman Images:** Photo © Boltin Picture Library. **34 The Metropolitan Museum of Art:** Gift of J. Pierpont Morgan, 1917. **35 Getty Images:** Archiv Gerstenberg / ullstein bild Dtl. / Contributor. **36 © Cartier:** Nils Herrmann, Collection Cartier. **37 Getty Images:** Arthur Edwards- WPA Pool. **40 Shutterstock.com:**

192

DK Secret Histories